INTRODUCTION TO SPACE SCIENCES

AND SPACECRAFT APPLICATIONS

INTRODUCTION TO SPACE SCIENCES

AND SPACECRAFT APPLICATIONS

BRUCE A. CAMPBELL AND SAMUEL WALTER MCCANDLESS, JR.

Gulf Publishing Company
Houston, Texas

Introduction to Space Sciences and Spacecraft Applications

Gulf Publishing Company
Book Division
P.O. Box 2608 □ Houston, Texas 77252-2608

10   9   8   7   6   5   4   3   2   1

Library of Congress Cataloging-in-Publication Data
Campbell, Bruce A., 1955–
    Introduction to space sciences and spacecraft applications / Bruce A. Campbell, Samuel Walter McCandless, Jr.
        p.   cm.
    Includes bibliographical references and index.
    ISBN 0-88415-411-4
    1. Astronautics.   2. Space vehicles.   I. McCandless, Samuel Walter. II. Title.
TL791.C36   1996
629.4—dc20   96-14936
                                                        CIP

Printed on acid-free paper (∞)

Contents

Acknowledgments

The authors express appreciation to all the colleagues and students whose reviews and comments helped to shape and refine this book.

About the Authors

Bruce A. Campbell is currently involved in spacecraft project management at NASA's Goddard Space Flight Center. He is an aerospace engineering graduate of the U.S. Naval Academy and flew carrier-based jet aircraft as a Naval Flight Officer.

Samuel Walter McCandless, Jr., is president and founder of User Systems, Inc., which provides airborne and space-based radar and microwave sensor and system design and evaluation, remote sensing and satellite system design, and associated services. He was test director for NASA's Surveyor 1 spacecraft and managed NASA's Seasat program from its inception through operation of the satellite in space.

Preface

In 1985, the authors were asked to create an introductory course for a new series of classes focusing on space systems engineering in an undergraduate aerospace engineering curriculum. Although the authors were familiar with separate texts on each of the subjects to be taught, there seemed to be no single existing text that covered all the topics at a level that could be used for an introductory class. The first time the course was taught, handouts were created for each of the topics covered. Each successive time that the course was taught, the handouts were updated based on the experience of the preceding class. The idea to publish came very early, and the handouts evolved into the chapters that make up this text.

In its manuscript form, this text has been used by several hundred students, in several different institutions, under different instructors, with very favorable reviews. The most common comment is the ease with which the student can read and understand the material. Although the text is geared toward sophomore/junior undergraduate engineering students who will continue to take courses in these space-related topics, it has proven to be comprehensible and interesting to students in other science and nonengineering fields as well. The subjects are introduced on a basic level with no prior related knowledge expected of the student. A fundamental knowledge of physics, differential equations, dynamics, and other pre-engineering subjects is helpful, but not necessary, to understanding the basic concepts presented.

To emphasize this point, the subject matter of the text has been condensed into a "short course" that has been presented to many diversified groups of managers, technicians, military personnel, and other professionals—not necessarily engineers—over the past several years. Many of these people are involved in space systems acquisition or operation, and most report that the course provides them with a more complete level of understanding that makes them more comfortable in their fields and which they believe will help them in their professions. The text material has proven to be an excellent reference and review source for the student and professional alike.

This text considers many of the interdisciplinary topics necessary for understanding the design and application of space-based systems. The *space science* topics include orbital mechanics (Chapter 2), propulsion (Chapter 3), and a description of the space environment (Chapter 4). These chapters prepare the way for discussing *spacecraft applications*. The most common uses of systems in space include communications (Chapter 5), remote sensing (Chapter 6), and navigation (Chapter 7). Finally, the basic systems required by most spacecraft (Chapter 8) and a methodology used to design a spacecraft (Chapter 9) are presented.

In a four-year, semester-oriented undergraduate institution, it is recommended that this text be used for a 3-credit-hour (2 hours classroom, 2 hours laboratory) course given at the end of the sophomore year or beginning of the junior year. By this time the student should be familiar with most of the helpful subjects discussed above and will be ready to take this course and, if desired, any of the follow-on space topic courses. Early in the course, the laboratory hours can be used for additional classroom hours and problem solving; however, as the course progresses, laboratory hours could include orbital mechanics computer simulations, Estes rocket firings, an "outside" environmental lab, a communications lab, review of some remote sensing materials or films, a GPS navigation demonstration, or other demonstrations that help the student get a feel for the subjects discussed. Later in the course, the laboratory hours should be devoted to the formulation of an idea for a space-based application and the top-level conceptual design of a system to fit the need.

This text could also be used for a 2- or 3-hour classroom-only course which might be more suitable to the nonengineering student or two-year institution.

CHAPTER 1

Introduction and History

To set the stage for what is presented in this text, a brief discussion on the uses of systems in space is provided as insight into the operation and purposes of such systems. This discussion is followed by a brief history of man's achievements in space and some conjecture on where we may be headed.

USES OF SPACE

For centuries, the stars and planets and their patterns and positions have been used for purposes similar to those for which we use modern space systems. Notable and ancient applications include navigation and environmental prediction. So it is surprising that even after Sir Isaac Newton showed that man could, theoretically, place an object into space, no one immediately thought of any useful reason why anybody would want to. It was not until this feat was actually accomplished that we began to explore and discover the benefits offered by presence in space and began designing and building systems to take advantage of this wonderful new dimension.

After the first few pictures were returned from some of the early suborbital test rocket flights, experimenters recognized that one of the greatest advantages of being in space was the expanded perspective looking back toward earth. The ability to view areas of the globe large enough to show entire storm patterns spawned the science of global environmental remote sensing. This has dramatically increased knowledge of the environment, allowing for the observance and prediction of adverse conditions and thereby saving life and property from disaster. Remote sensing systems also allow us to discover and monitor natural resources and evaluate demographic effects such as urban expansion and pollution. Similar type sensors, positioned outside the blanket of the earth's atmosphere and

1

on interplanetary craft, allow a clearer look at the universe of which we are such a small part, giving us a better understanding of where we have been and what the future may hold.

The need for advanced global communication capabilities caused many to look toward space as an alternative to earlier methods. As early as the 1940s, science writer Arthur C. Clarke hypothesized that a platform placed at a particular location in space (where it would appear motionless with respect to the ground) could be used as a communications relay station. A multitude of such systems exist today, and the resulting real-time, global availability of information has transformed the news, entertainment, and communications industries and has had a major impact on political and military operations as well.

When the first artificial satellite, the Soviet Union's *Sputnik 1,* began circling the globe, U.S. scientists discovered a method for using the signals transmitted by the satellite to accurately determine one's position anywhere on the earth. Satellite navigation is now employed by military, government, private, and commercial users for positioning and surveying on land as well as for aircraft and ship navigation.

When man himself ventured into space, many benefits resulted as byproducts of the developments in technology required to support these missions. Astronauts conducted experiments that proved that the unique environment of space offered many useful possibilities. In microgravity

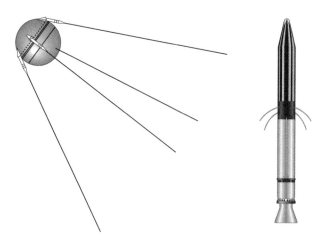

Figure 1-1. Sputnik (left) and Explorer (right)—the first satellites launched by the Soviet Union and the United States, respectively.

environments, materials processing impossible in the gravity-constrained laboratories on earth is possible and may result in new alloys, crystals, and medicines which could improve man's existence. The need for electrical power for space systems has lead to alternative sources of energy for use on earth. The future uses of space are still developing, and space may truly become mankind's next frontier.

HISTORY OF SPACEFLIGHT

Even though many consider the launch of *Sputnik 1* by the Soviet Union as heralding the Space Age, man's efforts toward this endeavor actually began much earlier. Notwithstanding the dreams of writers such as H. G. Wells and Jules Verne, the first practical efforts toward spaceflight really began with early rocket pioneers such as Russia's Konstantin Tsiolkovsky, Germany's Hermann Oberth, and America's Robert Goddard. The V-2 rocket, used to terrorize England during World War II, was the first operational rocket system. America's advances began after the war with the aid of expatriated German scientists, including Wernher von Braun, and a cache of captured V-2s with which to experiment. The Soviet Union received the same type of reparations after the war and began a similar effort.

The first efforts by both countries were in developing improvements to the V-2, and well before *Sputnik 1,* both countries had operational intercontinental ballistic missile (ICBM) systems. Many of the V-2 rockets and follow-on booster designs carried scientific experimental packages on board during test flights. These packages measured the characteristics of the atmosphere, the ionosphere, and the low-altitude space environment. One such package was used by Dr. James Van Allen to discover the existence of regions of highly energetic particles encircling the globe, subsequently named the *Van Allen radiation belts.*

The Soviet Union placed the first man-made object into orbit around the earth on October 4, 1957. The overhead signals received from the 83-kg (184-lb) *Sputnik 1* ("Traveler") alarmed many who demanded that the United States match the feat. Pressure increased when a month later *Sputnik 2,* at a surprising 507 kgs (1,120 lbs), went into orbit carrying a small dog to test the effects of space on a living creature. (The dog Laika survived for a week until purposely poisoned, because the craft was not capable of reentry.) America finally succeeded with *Explorer 1,* launched on January 31, 1958, but U.S. capabilities were severely questioned when comparing the 8-kg (18-lb) satellite to the two prior Sputniks.

Figure 1-2. Launch of Alan Shepard, America's first man in space, on the Mercury-Redstone-3 mission, May 6, 1961. *(Photograph courtesy of NASA.)*

In an effort to consolidate and accelerate America's efforts in space, Congress authorized the establishment of the National Aeronautics and Space Administration (NASA) in July 1958. The new agency absorbed the National Advisory Committee for Aeronautics (NACA) as well as many military space projects, including their personnel and facilities. Both countries continued to launch larger and more sophisticated unmanned satellites for a variety of military and scientific purposes, but it was apparent that efforts were also being made to place a man into space.

Manned Spaceflight

Once again, the Soviets were first in the manned spaceflight effort with the launch of Air Force Major Yuri A. Gagarin in the 4,706-kg (10,400-lb) *Vostok 1* ("East") spacecraft on April 12, 1961, for a single orbit. The recovery area, unlike the later U.S. manned flights, was back in the Soviet homeland, and, after the spacecraft reentered and was on its way down by parachute, Gagarin exited the capsule and used his own parachute to land! Three weeks later on May 5, 1961, the United States launched Navy Commander Alan B. Shepard, Jr., into space. The successful 15-minute suborbital flight reached an altitude of 187.6 km (116.5 mi) and went 489.1 km (303.8 mi) downrange to recovery in the South Atlantic. Although successful, this flight showed, once again, that America was lagging behind the Soviet Union in what was now being called the "Space Race." In an effort to turn this tide, President John F. Kennedy, during a major speech before Congress only three weeks later, publicly committed the United States to landing a man on the moon, and returning him safely to earth, before the decade was out. With this clear objective in mind, the U.S. manned space program took shape, evolving from the existing Mercury program, through Gemini, to Apollo.

Mercury. The purpose behind the Mercury program was to gain basic data on the effects of spaceflight on human beings. To gain preliminary physical data as well as test the launcher and spacecraft systems, a rhesus monkey (Sam, December 1959) and a chimpanzee (Ham, December 1960) were launched on suborbital flights, and another chimp (Enos, November 1961) was launched into orbit before these same flights were attempted by astronauts.

Alan Shepard's flight was followed by a second suborbital flight carrying Virgil "Gus" Grissom in July 1961. John Glenn became the first American to orbit the earth on February 20, 1962. Three additional orbital Mercury flights followed, the longest by Gordon Cooper who spent more than 34 hours alone in orbit.

Gemini. Gemini flights were designed to evaluate the ability of performing the tasks in space required for a manned lunar landing. Three methods

Figure 1-3. Photograph of the two-man *Gemini 7* spacecraft taken from the *Gemini 6* spacecraft during rendezvous maneuvers in December 1965. (*Photograph courtesy of NASA.*)

of conducting the moon flights were under consideration. One was the *direct ascent* of a spacecraft to the lunar surface and back. Another was known as *earth-orbital rendezvous,* by which two spacecraft would be launched separately to rendezvous in the earth's orbit before continuing as one to the moon and back. The method ultimately selected was called *lunar-orbit rendezvous,* by which a spacecraft would be launched into lunar orbit whereupon a specially designed lunar landing module would travel down to the surface and back up to lunar orbit to redock with the mother ship for the return to earth. In order to accomplish this, crew size would have to be increased, longer flight duration would be necessary, and in-orbit operations such as extravehicular activities (EVAs) and rendezvous and docking techniques would have to be perfected.

The Gemini capsule was basically a modified Mercury capsule, enlarged to carry two men, with the capability to perform the required on-orbit operations. There were ten Gemini flights in the two years the program ran from 1965 to 1966. These flights were a significant evolution from the earlier Mercury flights in which the astronauts were basically only along for the ride, and the outstanding successes of Gemini gave NASA confidence in its abilities to continue on toward the moon.

Apollo. Unlike the earlier flights which used modified ICBMs as launchers, Apollo required a totally new launch vehicle designed specifically to carry men and equipment to the moon. In 1963, even before the first Gemini flights, Wernher von Braun's team at the Marshall Space Flight Center in Huntsville, Alabama, was given the go-ahead to develop his Saturn family of powerful boosters. The Saturn V rocket still represents the most powerful launch vehicle ever developed in the United States and was a "giant leap" in technology at the time.

However, the confidence gained from the Gemini flights was shaken in January 1967 when a fire broke out in the *Apollo 1* capsule during a prelaunch test, killing astronauts Gus Grissom, Ed White, and Roger Chaffee. Redesign of the capsule to prevent any similar occurrences delayed the first manned Apollo flight, designated *Apollo 7,* until October 1968.

Only nine months later, on July 20, 1969, astronauts Neil Armstrong and Edwin "Buzz" Aldrin entered the gangly-looking lunar excursion module (LEM) and descended to the surface of the moon while Mike Collins orbited overhead in the *Apollo 11* command module. Between this date and December 1972, five more crews landed on the moon. The twelve astronauts deployed experiments, explored the surface, and returned a total of 378 kgs (836 lbs) of lunar material for study by the international scientific community.

The only other mishap occurred on *Apollo 13,* when an oxygen tank in the service module section of the spacecraft exploded on the way to the moon. The tense situation had a happy ending when the crew was recovered six days after launch, having used the lunar module as a lifeboat and booster rocket to push themselves around the moon and back toward the earth in place of the ailing command/service module.

The Apollo program had successfully met its objectives, and the United States had clearly accomplished an enormous feat. But what next?

Skylab. After the lunar landings, public and governmental support for the space program diminished. With the escalation of the Vietnam War, people began to question the relative worth of the 25 billion dollars spent for the program. Three planned lunar landings (Apollo 18, 19, and 20) were canceled as a result of a review of the nation's space program by a Space

Figure 1-4. *Apollo 15* astronaut on the surface of the moon with the lunar excursion module and lunar rover vehicles nearby. (*Photograph courtesy of NASA.*)

Task Group of the National Academy of Sciences established by President Richard M. Nixon in 1969. The new focus would be on development of a reusable space transportation system.

Initial plans also called for two separate space stations to be visited by multiple crews using Apollo-derived equipment. Instead, a single station, Skylab, was approved. When it was launched atop the last Saturn V booster in May 1973, a meteoroid shield and one of the two main solar panels were lost during ascent. The first crew to arrive found living conditions unbearably hot and a station lacking power. Through ingenuity and unplanned EVAs, the crew was able to erect a sunshield and straighten out the power problem to make the station usable. The longest mission to *Skylab* was by the third and final crew in 1974 and lasted 84 days.

Skylab was used for a wide variety of investigations, including pioneering earth and atmosphere remote sensing, materials processing, biomedicine, and some of the most revealing solar studies to date. Abandoned after a total of 171 days in use, *Skylab*'s orbit gradually decayed until it finally reentered the atmosphere in July 1979 and was destroyed.

Apollo-Soyuz. As an act of détente, the United States and the Soviet Union agreed to a joint space operation involving the docking of an Apollo capsule with a Soyuz spacecraft. Scientists, engineers, and crews from both countries worked together to design a universal docking adapter to enable the two ships to dock and give direct access between the spacecraft. Both craft were launched on the same day in July 1975 and docked two days later. In addition to ceremonies and interviews for television, the crews conducted joint experiments using facilities on board both spacecraft. Though not technologically critical, the Apollo-Soyuz mission demonstrated that the two superpowers could cooperate in the peaceful use of outer space.

Space Shuttle. From 1975 to 1981, no American manned missions were launched. NASA focused on the development of the reusable space transportation system (STS) commonly referred to as the Space Shuttle. The

Figure 1-5. Three crews visited *Skylab* between May 1973 and February 1974. (*Photograph courtesy of NASA.*)

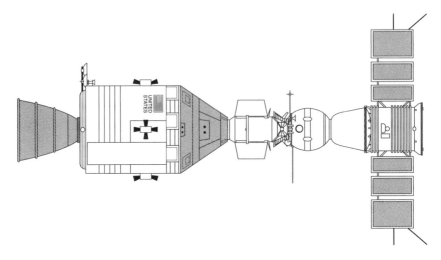

Figure 1-6. Apollo/Soyuz: The United States and Soviet spacecraft remained docked for two days while the crews performed joint experiments.

shuttle was seen as the next step in spaceflight evolution, allowing routine access to space. New satellite designs would take advantage of the large payload compartment and relatively easy trip into orbit compared to conventional launchers. Satellites and experiments could be deployed, serviced and repaired, retrieved, and returned to earth by the shuttle. The flight crew, which would include nonpilot scientists and technicians, could conduct experiments and observations in a shirt-sleeve environment or don space suits to conduct activities outside.

In its original conception, the Space Shuttle was to take off and land at conventional runways and be completely reusable for up to 100 flights, at least halving the cost per pound of carrying payloads into space. However, design difficulties and budgetary constraints lead to a simpler and less expensive configuration using solid rocket motors and an expendable external fuel tank to boost an orbiter into space which, after its mission was complete, would glide back to earth to land like an aircraft.

On April 12, 1981, the first space shuttle, *Columbia,* lifted off from Kennedy Space Center for a completely successful two-day mission with astronauts John Young and Robert Crippen on board. In a little under five years, the four-ship shuttle fleet flew 24 successful missions carrying satellites, experiments, and up to a seven-person crew into low-earth orbit. However, the twenty-fifth mission, launched on January 28, 1986,

Figure 1-7. The Space Shuttle can retrieve and return payloads as well as carry them, and astronauts, into low-earth orbit. (*Photograph courtesy of NASA.*)

turned into a disaster when hot gasses from a leak in one of the solid rocket booster segment seals ignited the fuel in the huge external tank. The subsequent explosion disintegrated the shuttle *Challenger,* killing the seven-person crew. Redesign of the booster seals and other shuttle systems delayed the next STS launch until September 1988.

An after-effect of the *Challenger* disaster was the decision to eliminate use of the shuttle to deploy commercial spacecraft, limiting flights to deployment of military, national, and scientific spacecraft which require the special capabilities of the shuttle, and construction of the planned space station. Despite this self-imposed limitation, the four-ship shuttle fleet has averaged about seven flights per year since resumption of operations.

Soviet Manned Spaceflights. Soviet and American space programs paralleled each other in many ways. The one-man Vostok spacecraft, the last of which carried Valentina Tereshkova, the first woman in space, was modified to carry a larger crew and was redesignated Voskhod ("Sunrise"). The Voskhod spacecraft was only used twice but carried a (cramped) crew of three on the first flight and was the test-bed for the world's first EVA on the second.

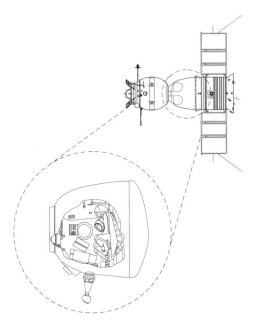

Figure 1-8. Only the Soyuz crew compartment is recovered after a flight.

Over two years elapsed before the next Soviet manned launch, which was the first flight of the new Soyuz ("Union") spacecraft only three months after the *Apollo 1* fire. This first flight ended in tragedy when the recovery parachute lines became tangled on reentry and the lone cosmonaut was killed on impact. Like Apollo, the Soyuz program was set back for more than a year before the next manned flight.

Evidence that the Soviets were working toward a manned landing on the moon came when a spacecraft, launched in 1971, reentered and crashed in Australia in 1981. Concern over an earlier reentry of a radioactive nuclear power generator forced the Soviets to reveal that the mission had been a test of a lunar cabin. Lack of a launch vehicle with sufficient thrust seems to be the reason for the slow Soviet progress in this effort, which was then abandoned after the U.S. successes.

From 1971 to 1982, six successful Salyut space stations were launched and visited by a total of 29 crews. About one-quarter the size of *Skylab,* the stations were used for a number of science, biomedical, and military missions, the longest lasting 237 days aboard *Salyut 7* in 1984. The following space station design, *Mir* ("Peace") was a step forward with its capability for longer missions and expansion and modification. Crews have remained on board for up to a year, and several docking and laboratory modules have been added to the basic core.

The Russians continue to use the Soyuz spacecraft (with modifications) to this day, concentrating on earth-orbit operations. For the past several years they have averaged two to three Soyuz flights per year, exclusively associated with visits to the *Mir* space station.

International Manned Spaceflight. The United States and Russia remain the only countries capable of launching and supporting manned spaceflight missions independently. However, individuals from several other countries have participated in various U.S. and Soviet/Russian manned missions. The Space Shuttle has carried several international payloads accompanied by payload specialists and scientists from these host countries. Besides cosmonauts from the former Soviet Union states, the Intercosmos program carried "guest" cosmonauts from other countries into space aboard the Soyuz capsule. Countries that have participated in these programs include Afghanistan, Austria, Bulgaria, Canada, Cuba,

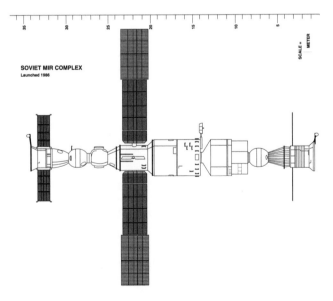

Figure 1-9. The *Mir* space station has been in near-continuous use since its launch in February 1986.

Czechoslovakia, France, Germany, Hungary, India, Japan, Mexico, Mongolia, the Netherlands, Poland, Romania, Saudi Arabia, Syria, the United Kingdom, Vietnam, and the United States.

A listing of manned spaceflight missions, crews, and mission highlights through the first flight of the Space Shuttle is given in Appendix A.

Unmanned Spaceflight

In addition to manned spacecraft, the United States, the Soviet Union/ Russia, and, more recently, many other countries have developed, launched, and operated hundreds of unmanned spacecraft designed to perform military, national interest, scientific, and commercial applications. The following sections highlight some of the application areas in which these spacecraft perform and the evolution of spacecraft within each area.

Earth Observation. Due to the unique vantage point of space, many satellites are designed to observe the earth to collect data on the environment. *TIROS 1* (Television Infrared Observation Satellite), the first weather satellite, was launched in April 1960 and returned over 22,500 pictures of the earth. TIROS, NOAA (National Oceanographic and Atmospheric Administration), GOES (Geostationary Operational Environmental Satellite), and the military DMSP (Defense Meteorological Satellite Program) systems monitor the earth's weather, allowing timely prediction of climatic changes and their effects on the planet.

In September 1991, the Upper Atmosphere Research Satellite (UARS) was deployed from the Space Shuttle to conduct the first comprehensive studies of the stratosphere, mesosphere, and thermosphere regions of the atmosphere. The *total ozone mapping spectrometer* (TOMS), a NASA instrument that "hitched a ride" on U.S. Nimbus and Soviet/Russian Meteor weather satellites, discovered and continues to monitor the "ozone hole" region of dangerously depleted ozone levels which appears over Antarctica each fall. Missions such as these allow researchers to map the short- and long-term changes occurring in the atmosphere and to determine what effects may be attributable to humans.

ERTS 1 (Earth Resources Technology Satellite) was launched in July 1972 into an orbit over the poles and used high-resolution electronic cameras to collect imagery on geology, crops, population, and pollution. ERTS satellites evolved into the Landsat series which continues to focus

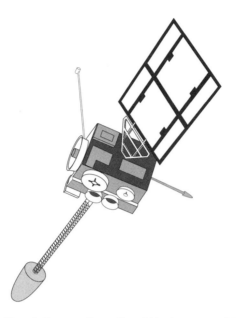

Figure 1-10. The Geostationary Operational Environmental Satellite (GOES).

on discovering and monitoring earth's resources in even the remotest areas of the world.

In June 1978, *Seasat* was launched to gather information on sea temperature, sea ice, wind speed and direction, and wave heights. Though its mission was cut short due to an electrical failure, *Seasat* first demonstrated the beneficial use of *synthetic aperture radar* (SAR) to provide all-weather, day and night, high-resolution images for a variety of earth-monitoring applications. A U.S./French satellite, *TOPEX/Poseidon,* launched in August 1992, uses a radar altimeter to precisely measure the shape of the earth (geodesy) as well as the changes in global sea states to help determine the ocean's role in the earth's climate.

Space Environment. James Van Allen used data from America's very first satellite (*Explorer 1*) to discover the "belts" of energetic particles encircling our world and which now bear his name. Subsequent Explorer spacecraft allowed scientists to probe the solar wind and the earth's magnetic field. The interaction between these phenomena was examined by the International Sun-Earth Explorer (ISEE) satellites, launched in 1977 and 1978, two in orbit around the earth and one positioned in a special orbit between the sun and the earth. Pioneer and Voyager spacecraft, sent to visit the planets, also sampled the interplanetary environment. These satellites are now on their way out of our solar system, but scientists continue to monitor their signals in hopes of gaining information on interstellar space.

The sun has the greatest effect on the near-earth space environment, and a series of Orbiting Solar Observatory (OSO) spacecraft launched between 1962 and 1975 were used to study solar flares and temperature differences on the sun's surface. The Solar Maximum Mission (SMM or *SolarMax*) was launched in 1980 to observe the sun during the maximum phase of its 11-year cycle. This spacecraft was the objective of a Space Shuttle rendezvous and EVA repair mission in 1984. *Ulysses,* a joint U.S./European Space Agency spacecraft deployed from the Space Shuttle in October 1990, used a gravity assist from Jupiter to send it into a solar-polar orbit to allow mapping the solar environment in this previously unexplored region of our solar system.

Long-term effects of the space environment on spacecraft has been investigated with the Long Duration Exposure Facility (LDEF). Released from the

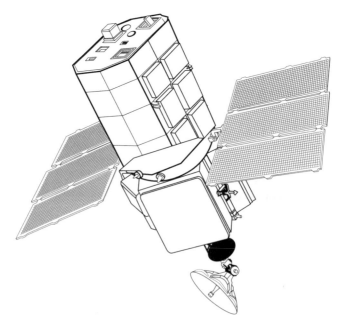

Figure 1-11. The *SolarMax* sun-observing satellite was the object of a Space Shuttle repair mission in 1984.

Space Shuttle in 1984 for a one-year stay, the *Challenger* accident delayed its recovery until 1990 when it was finally returned to earth for analysis.

Planetary Exploration. Planetary exploration began with flights to our nearest neighbor, the moon. Before sending men there, several Ranger spacecraft were sent crashing into the lunar surface, sending back high-resolution photographs right up until impact. Lunar Orbiters photographed the moon from orbit to locate possible landing sites, and Surveyor craft soft-landed on the moon and sent back photos and conducted lunar soil sampling experiments.

Mariner 2 became the first spacecraft to fly by another planet, passing within 21,000 miles of Venus in December 1962. Subsequent Mariner and Pioneer spacecraft were sent toward Mercury, Venus, Mars, Jupiter, and Saturn—all the planets known to early astronomers! In 1976, two Viking spacecraft successfully landed on Mars, sending back pictures of the Martian landscape and conducting soil and atmospheric experiments. Spectacular pictures and a wealth of planetary information on Jupiter, Saturn, Uranus, and Neptune were sent back from the two Voyager spacecraft, now on their way out of our solar system. The *Magellan* spacecraft used synthetic aperture radar (SAR) to map more than 98 percent of the surface of the cloud-enshrouded planet Venus from September 1990 to September 1992. *Galileo,* deployed in October 1989 from the Space Shuttle, used

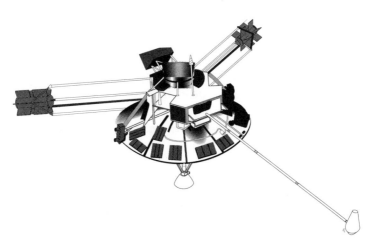

Figure 1-12. The Pioneer spacecraft made the first "up close" observations of many of the planets.

both Venus and the earth for gravity assist boosts to reach the planet Jupiter and release a probe into its atmosphere. Along the way, *Galileo* flew by the asteroids Gaspra (October 1991) and Ida (August 1993), giving us our first close looks at these solar system companions.

The Soviets sent numerous spacecraft to explore Venus, and two of their Venera craft successfully landed on the surface in March 1982. The landers were only able to transmit pictures and data for a short time until the harsh environment (temperatures which will melt lead and pressures about 100 times that on earth) disabled the spacecraft. While the United States conducted the first flyby of a comet (Giaccobini-Zinner) in 1985, a flotilla of spacecraft from the Soviet Union, the European Space Agency, and Japan made somewhat more noteworthy flybys of Halley's comet in 1987.

Space Exploration. An Orbiting Astronomical Observatory (OAO 2) was launched in December 1968 with a complement of ultraviolet instruments designed to examine the stars, and discovered evidence of the existence of "black holes." High-Energy Astronomy Observatory (HEAO) spacecraft were launched between 1977 and 1979 to conduct X-ray, gamma ray, and cosmic ray surveys of the universe, and the International Ultraviolet Explorer (IUE) conducted observations in the UV wavelengths. Launched in November 1989, the Cosmic Background Explorer (COBE) carried instruments designed to search the infrared background of deep space for information about the "Big Bang" theory of the origin of our universe.

Deployed in April 1990 from the Space Shuttle, the Hubble Space Telescope (HST) was the first of the "great observatories," facility-class satellites designed to offer scientists unparalleled opportunities to explore the universe. HST has a 2.4-meter (94.5-inch) diameter mirror which can pick up objects 50 times fainter and 7 times more distant than present ground-based observatories, potentially expanding the universe visible to astronomers by five hundredfold! Designed for on-orbit servicing, in December 1993 astronauts were able to insert a set of corrective optics to offset the "spherical aberration" flaw discovered in the primary mirror that compromised HST's capabilities. The second great observatory, the Compton Gamma Ray Observatory (GRO), was deployed from the Space Shuttle in April 1992. GRO was designed to capture the relatively unimpeded gamma ray emissions from cataclysmic cosmic events like supernovas, black holes, and even the remnants of the Big Bang. Other great observatory satellites may follow.

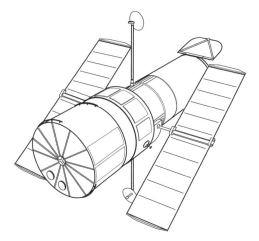

Figure 1-13. Hubble Space Telescope. The HST was the object of a space shuttle repair mission in 1993.

Commercial Satellites. By far, the largest commercial interest in space systems is in the relay of information, or communications. Communication satellites had a humble start with *Echo 1,* launched in August 1960, which was simply a huge radio-reflecting balloon in low-earth orbit. By 1963, *Syncom 1* demonstrated the capability and advantages of communicating via satellite in a *geostationary* orbit. Today, over one hundred communication satellites exist in this orbit, each capable of relaying thousands of telephone calls and television pictures simultaneously around the globe. In 1962, President Kennedy signed a bill creating the Communications Satellite Corporation (Comsat), a joint government-industry undertaking allowing transfer of U.S.-developed satellite communications technology for commercial use. The International Telecommunications Satellite Consortium (Intelsat), with Comsat representing the United States, promotes international use of these capabilities. Both organizations have been highly successful.

A similar arrangement exists involving dissemination of data gathered by the Landsat satellites for which demand by the commercial public has grown. The French SPOT (Satellite Probatoire de l'Observation de la Terre) satellite competes in this area, offering high resolution photographs of points of interest on the earth to news media, governments, and even military organizations worldwide.

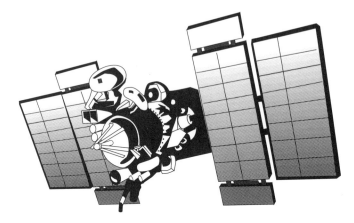

Figure 1-14. Gorizont. Commercial channels have been made available on this Russian geosynchronous communications satellite.

Military Satellites. The early experiments using captured V-2 rockets were conducted under the auspices of the military and led to the development of the intercontinental ballistic missile (ICBM). As other uses of space became apparent, the military conducted their own experimental programs and developed systems to take advantage of the new "high ground" of space.

Today, the military routinely conducts force enhancement missions such as communications, navigation, and remote environmental sensing using dedicated systems in space. Surveillance and monitoring missions are also routinely conducted, though the systems used are often classified. However, the worth of these systems became generally apparent during the war with Iraq, where coalition forces scored a decisive victory using these capabilities heavily.

International Use of Space

Although the foregoing pages emphasize U.S. accomplishments and activities in space, Russia, Japan, France, and China all have fully developed programs capable of providing spacecraft manufacture, launch services, and operation of payloads. The scientific, environmental sensing, and commercial communications use of space systems is an international arena. The United States and Russia no longer dominate the field, with France, China, and others vying for commercial customers for space sys-

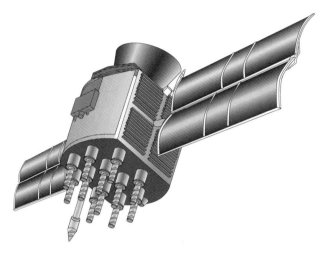

Figure 1-15. Navstar/GPS. Though designed to provide precision navigation for military forces worldwide, commercial use of the Global Positioning Satellite system has become widespread.

tems and services. Nations such as the United Kingdom, Canada, Italy, India, Korea, and others have a presence in space with specific satellites or technologies geared to space applications. Space has become a worldwide area of interest and commerce with many nations joining together to pursue mutual interests. The European Space Agency (ESA), a consortium of space organizations from 13 countries, is one such example. A listing of many U.S. and foreign satellites is given in Appendix B.

The Future

While the Space Shuttle will continue to be used to deliver people and material to outer space, future plans call for a more robust mix of launch vehicles which may include an increasingly wider range of expendable launch vehicles (ELVs) or more reusable concepts such as follow-on space shuttle designs, single-stage-to-orbit (SSTO) vehicles, or a National Aerospace Plane (NASP) capable of reaching space using more conventional take-off and landing operations. The Russians have tested a shuttle-like craft, *Buran* ("Snowstorm"), although the likelihood of its further development and use is doubtful. The Russians also have developed a heavy-lift booster of the Saturn V class, *Energia,* which could be used to launch complete space station modules into low-earth orbit. The European Space Agency continues to upgrade the Ariane booster and provide commercial launch services to customers worldwide.

The Russians continue their exploitation of space through continued use and expansion of the *Mir* space station, which includes long-duration missions of American astronauts and visits of the U.S. Space Shuttle. The Russians will also contribute to the international space station effort along with the United States, Canada, Japan, and the European Space Agency. Uses of the space station will include biomedical and material processing experiments conducted within the laboratory modules, as well as externally-placed earth-observing and astronomical instruments.

The Ballistic Missile Defense Office (BMDO—formerly the Strategic Defense Initiative, or SDI) envisions a missile defense system which includes space-borne assets. Success in the Persian Gulf War ensures continuance and updating of military satellite systems including the Global Positioning System (GPS) for navigation, the MILSTAR and other net-

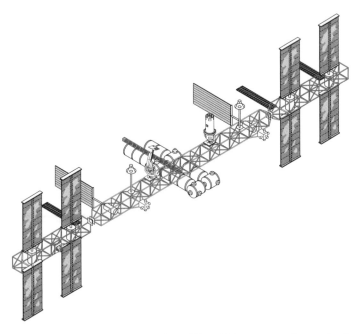

Figure 1-16. International space station. This 1993 concept shows the basic building blocks of the planned space station: habitation and laboratory modules supported by solar power arrays and other systems on a long truss structure.

works for communications, and other military systems for weather and surveillance uses.

NASA's Mission *to* Planet Earth is an effort to characterize the earth's environment and the changes it may be undergoing. This effort includes the Earth Observing System (EOS), a series of large "observatory"-sized satellites designed for simultaneous measurements of the earth's environment from space, as well as several smaller specialized satellites and instruments from many countries.

NASA's Mission *from* Planet Earth looks outward to man's continued exploration of the universe. It includes the moon/Mars initiative which promotes a lunar base and a manned mission to Mars, possibly as an internationally cooperative effort. This initiative includes unmanned missions to these and other bodies in our solar system and provides an impetus for increased development and use of robotic systems.

It is obviously man's intention to expand his presence and capabilities in this frontier, and the future of space exploration, no matter how unpredictable, will certainly be challenging, exciting, and international in character.

REFERENCES/ADDITIONAL READING

Kerrod, R., *The Illustrated History of NASA,* Anniversary Edition. New York: Gallery Books, 1988.

Bilstein, R., *Orders of Magnitude: A History of the NACA and NASA, 1915–1990.* Washington, D.C.: U.S. Government Printing Office, 1989.

Clark, P., *The Soviet Manned Space Program.* New York: Orion Books, 1988.

Simpson, T. (Ed.), *The Space Station.* New York: IEEE Press, 1985.

Thompson, T. (Ed.), *TRW Space Log, Vol. 27, 1957–1991.* Redondo Beach: TRW Space & Technology Group, 1992.

EXERCISES

1. List and expound upon some of the benefits achieved by using remote sensors in space.

2. Repeat Exercise 1 for satellite navigation systems.

3. Repeat Exercise 1 for satellite communications systems.

4. Expound upon the beneficial uses of the above systems to units of the various military services (Army, Navy, Air Force, Marines, Coast Guard).

5. List any other applications you can think of for using systems in space.

6. Repeat Exercise 1 for one of these other applications.

7. Discuss the differences in the approaches taken in space by the United States and the Soviet Union (Russia/Commonwealth of Independent States).

8. What do you think the ramifications of other countries (or groups of countries) establishing an independent space capability might be?

9. What, in your opinion, should be the next major effort in man's venture into space (e.g., space station, moon base, Mars, commercialization) and why?

10. What are your opinions about international cooperation in these efforts?

PART 1

Space Sciences

CHAPTER 2

Orbital Principles

Around 350 B.C., the renowned Greek philosopher Aristotle made the deduction that was to influence man's perception and understanding of the motions of the planets for over two thousand years! Aristotle was of the opinion that all the heavenly bodies, including the sun, the wanderers (planets), and the stars, were in circular motion about a fixed and unmoving Earth. So authoritative was Aristotle that the Catholic church later accepted this view as its official doctrine on the matter and used its religious influence to enforce this position. As a direct result, Galileo was arrested for espousing views to the contrary in 1600 A.D.! However, a century before Galileo, the first serious questioning of Aristotle's views had already begun.

Astronomy (or rather astrology) was very popular at the time of Copernicus, around 1500 A.D., and the inability to accurately describe the motions of the heavenly bodies increased efforts to find a better explanation than Aristotle's. Copernicus was both an astronomer and mathematician, and he used the observed angular positions of the planets and trigonometry to correctly place the solar system in its proper order, with the sun at the center and the earth as just another wanderer in motion about the sun like the other planets. However, the idea of a moving earth and similarity with the other planets was quite radical at the time and made his *heliocentric* (sun-centered) hypothesis hard to accept.

For instance, Tycho Brahe, the late sixteenth-century astronomer, completely rejected the notion. Tycho, in the years prior to his death in 1601, conducted the most exhaustive and accurate recording of the movements of the planets to date. He was sure that his data held the secret to the mystery of planetary movements, but his mathematical ability was too poor to check out his theories. Therefore, Tycho solicited the assistance of mathematicians like Johannes Kepler.

After Tycho's death, Kepler came into possession of most of the observational records kept over the many years. Kepler believed in the Copernican heliocentric hypothesis but was still saddled with the religious-backed belief in the perfection of the heavens, which called for purely circular motions of the planets. For nine years, Kepler struggled to fit the observed motions of the planet Mars into different models of combinations of circular paths. It was only after finally trying an oval-shaped path that Kepler found that the orbit of Mars fit extremely well into an ellipse. Kepler summarized his findings in his famous three "laws" which will be examined in this chapter. Kepler showed that these relationships held not only for the paths of the planets around the sun, but for the paths of the four moons recently discovered circling the planet Jupiter by Galileo. We will use these same relations to describe the motion of earth-orbiting satellites.

Galileo Galilei was well known, even in his own time, as a capable scientist and experimenter in many fields. In his quest for knowledge, Galileo employed a systematic approach to his studies that laid the foundation for what is now known as the *scientific method* of investigations. His study of the motions of bodies led to the understanding of friction, inertia, and the acceleration of falling bodies. Galileo was the first to use a combination of lenses to make telescopic observations, and with these crude devices he made some astounding discoveries, including the existence of the moons around Jupiter now known as the Galilean moons. It was the combination of Galileo's experiments in general motions and Kepler's findings in planetary motions that gave Newton the tools he needed to *explain,* not just describe, the mechanics of orbits.

Sir Isaac Newton's status as a genius is obvious considering the great leap of understanding in many different areas attributed to the seventeenth-century physicist. By his time, people had had over 50 years to mull over the findings of Kepler and others concerning planetary motions. It is important to recognize that Kepler's "laws" were derived empirically, simply describing the characteristics of the planetary motion recorded by Brahe. Yet no one had come up with a plausible explanation as to *why* the planets followed the particular paths around the sun that Kepler's relationships indicated, and there was great debate among the scientific investigators of the time. Sir Edmund Halley, a friend of Newton's, nonchalantly mused on the topic one day, and to his amazement Newton replied that he had come up with an explanation some 20 years earlier but had not bothered to publish the results. At Halley's urging and expense, Newton published his findings which included his three laws of motion and his description of gravitational force.

Newton and Kepler formulated their "laws" based on the motions of the planets around the sun, but these relationships describe orbital motions between any two bodies in the universe! In the following sections we will look at these relationships and how they describe the motions and other characteristics of bodies in orbits around the earth.

ORBITAL PRINCIPLES

Kepler's Laws

Kepler's First Law. Kepler's first law reveals that "captive" satellites (those with closed orbital paths) will travel around the earth in elliptical (or circular) paths with the center of the earth located at one of the foci, as depicted in Figure 2-1.

Orbital Parameters. Figure 2-1 defines some of the terms with which we will be dealing, and the geometry of the ellipse reveals some useful relationships between these orbital parameters. From the figure it is obvious that:

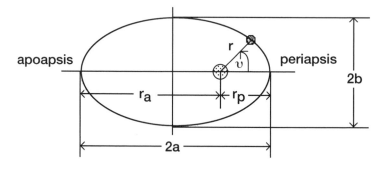

    a = semi-major axis   b = semi-minor axis
      Note: a circle is simply an ellipse with a = b
    r = radial distance between bodies' mass centers
    r_a = apoapsis radius (maximum distance between bodies)
    r_p = periapsis radius (minimum distance between bodies)
    υ = true anomaly (measured in same direction as movement in the orbit)

Figure 2-1. The ellipse. Many orbital terms can be defined simply by the geometry of an ellipse.

$$r_a + r_p = 2a \quad \text{or} \quad a = \frac{r_a + r_p}{2} \tag{2-1}$$

where r_a and r_p represent the *apoapsis* (maximum) and *periapsis* (minimum) distances between the bodies, respectively. These distances will be better defined shortly. The quantity **a** is defined as the *semi-major axis* of the ellipse.

The *eccentricity* **e** describes the shape of an ellipse, sort of in terms of how fat to wide with respect to the semi-major axis and semi-minor axis **b:**

$$e = \sqrt{1 - \frac{b^2}{a^2}} \tag{2-2}$$

A more useful form for the eccentricity equation can be derived from geometry as:

$$e = \frac{r_a - r_p}{r_a + r_p} \tag{2-3}$$

The orbital parameters **a** and **e** together define the *size* and *shape* of an ellipse, as exemplified in Figure 2-2 in which both ellipses have the same semi-major axis but different eccentricities.

The general equation for a conic section, of which an ellipse is just one type, may be written in a form which reveals many useful relationships:

$$r = \frac{a(1 - e^2)}{1 + e(\cos \upsilon)} \tag{2-4}$$

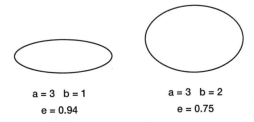

a = 3 b = 1   a = 3 b = 2
e = 0.94   e = 0.75

Figure 2-2. Eccentricity defines the shape of an ellipse.

In an orbit, **r** represents the *radial distance* (orbital radius) between the bodies' *mass centers*. The *true anomaly* υ is the angle measured from the major axis line (in the direction pointing toward periapsis) to the radial line between the two bodies. This is measured in the same direction as the motion of the orbiting body (refer to Figure 2-1).

Example Problem:

With the knowledge (from geometry) that for a circle the distances **a** and **b** are equal, find the eccentricity of a circular orbit and give the relationship between the periapsis, apoapsis, and semi-major axis distances.

Solution:

From equation 2-2 with **a = b**, it is clearly seen that *the eccentricity of any circular orbit is zero.*

Relating this finding to equation 2-3, for this expression to be zero, $\mathbf{r_a}$ must be equal to $\mathbf{r_p}$.

In fact, inserting a value of zero for eccentricity into the conic section equation (eq. 2-4) we find that **r = a** for *any* true anomaly υ. This result tells us that the *orbital radius for any circular orbit remains constant throughout the orbit.*

Orbital Types. A closer look at the conic section equation (eq. 2-4) reveals that the eccentricity can tell us immediately what *type* of orbit we are in:

- If we only consider positive eccentricities, the first limiting case is when **e = 0**, which we already know indicates a circular conic section and thus, a *circular* orbit.
- The next limiting case occurs when **e = 1**. For this value the conic equation gives a value of infinity for the radius at the point where the true anomaly approaches 180°. This corresponds to a *parabolic* conic section. Values of eccentricity greater than 1 indicate a *hyperbolic* conic section. Parabolic and hyperbolic orbits represent "open" or non-repeating orbits. Such orbits are used by deep space probes to escape the earth or even the solar system as in the case of the Voyager probes.

- The final case, then, is when the eccentricity is greater than zero but still less than one (**0 < e < 1**). For fixed values of the eccentricity and semi-major axis, we can see that the orbital radius is a function of the true anomaly which is a measure of exactly *where* along the conic section the orbiting body is. Substituting different values for the true anomaly, one would find that the minimum radius occurs when $\upsilon = \mathbf{0°}$ and the maximum radius occurs when $\upsilon = \mathbf{180°}$. These distances correspond to the *periapsis* and *apoapsis* points in the orbit shown in Figure 2-1 and prove that they are associated with the minimum and maximum distances between the bodies in an elliptical orbit. Values of eccentricity within this range, then, correspond to an elliptical conic section and would indicate an *elliptical* orbit.

At periapsis ($\upsilon = 0°$) and at apoapsis ($\upsilon = 180°$), the conic section equation (eq. 2-4) simplifies to reveal two useful relationships:

$$r_a = a(1 + e) \quad \text{and} \quad r_p = a(1 - e) \tag{2-5}$$

It is important to realize that orbital radius increases *continuously* from periapsis to apoapsis, and decreases continuously from apoapsis to periapsis when moving in an elliptical orbit.

Kepler's Second Law. Kepler's second law reveals that a line drawn between the two bodies will sweep out the same amount of area during the same time period anywhere along the orbital path. This characteristic is illustrated in Figure 2-3.

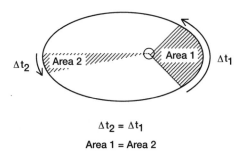

$$\Delta t_2 = \Delta t_1$$

Area 1 = Area 2

Figure 2-3. Area/time relationship. Kepler's second law reveals that the velocity of an object changes with orbital radius.

The most important point this law reveals is that the speed of an object in an orbit changes with changing distance between the bodies. For the areas shown in Figure 2-3 to be the same, the orbiting body must slow down when it is farther away so that the line between the bodies sweeps out the same area in the same amount of time as when it is traveling closer (both Δt's shown are equal). When Newton's laws are discussed, we will find a useful expression which shows this dependency of velocity on orbital distance.

Kepler's Third Law. In his third law, Kepler reveals that a relationship exists between the semi-major axis and the period of an orbit. Stated mathematically:

$$T^2 = \left(\frac{4\pi^2}{\mu} \right) a^3 \qquad\qquad (2\text{-}6)$$

In equation 2-6, **T** represents the *orbital period,* or time it takes to travel through one complete orbit with semi-major axis **a.** The term μ represents a *gravitational parameter* which has a specific value for each body around which an orbit may be described. The gravitational parameter for the earth is:

μ	English units	Metric units
Earth	1.4077×10^{16} ft^3/sec^2	3.986×10^5 km^3/sec^2

Use of this form of the gravitational parameter in Kepler's time equation (eq. 2-6) with the proper units used for the semi-major axis results in an orbital period given in seconds.

Example Problem:

With the knowledge that it takes the moon 27.32 (sidereal) days to complete one orbit around the earth, determine the semi-major axis of the moon's orbit.

Note: a *sidereal* day is the time it takes for the earth to complete one full rotation about its axis with respect to the (inertially fixed) stars. A *solar* day, measured with respect to the sun, differs due to the fact that the earth is also in motion around the sun.

1 sidereal day = 23 hrs, 56 min, 4 sec = 86,164 sec

Solution:

$$T = (27.32 \text{ days}) \times (86{,}164 \text{ sec/day}) = 2{,}354{,}000 \text{ sec}$$

Solving equation 2-6 for the semi-major axis:

$$a^3 = \left(\frac{\mu}{4\pi^2}\right) T^2 \qquad\qquad (2\text{-}7)$$

we find that for the period computed above:

$$a = 1.26 \times 10^9 \text{ ft} = 237{,}658.8 \text{ mi (using 5{,}280 ft/mi)}$$
$$= 382{,}469.6 \text{ km (using the metric values)}$$

The semi-major axis represents the *mean distance* of an orbiting body from the center of the earth. If the moon's orbit were circular, this distance from the earth would remain constant (as we have already shown). In fact, the moon's orbit is slightly elliptical, so its distance from the earth is constantly changing as the following problem shows.

Note: the apoapsis and periapsis points of an orbit described around the earth are called *apogee* and *perigee* respectively.

Example Problem:

The moon is in a slightly elliptical orbit around the earth (e = 0.055). Determine the moon's apogee and perigee distances in both miles and kilometers.

Solution:

Recalling the values obtained earlier for the moon's semi-major axis, and using the relationships obtained from the conic equation at apoapsis and periapsis (eq. 2-5):

$$r_a = a(1 + e) = 250{,}730.0 \text{ mi} = 403{,}505.4 \text{ km}$$

$$r_p = a(1 - e) = 224{,}587.6 \text{ mi} = 361{,}433.7 \text{ km}$$

This represents an appreciable change in distance for an orbit considered almost circular and would certainly have to be taken into account if planning a trip to the moon.

Newton's Laws

Kepler's laws were empirically derived relationships describing planetary motions based on years of observations. Newton's laws more precisely defined the mechanics of motions and, combined with his idea of *universal gravitation,* allowed a more complete description of general orbital motions. In a dedicated course in orbital mechanics, Newton's laws of motions are studied in great depth, and in fact, Kepler's laws are usually derived from these relationships. For our purposes, the next few sections will merely state some of the important results of Newton's laws.

Angular Momentum. The *angular momentum* of an orbit is represented by a *vector* quantity perpendicular to the orbital plane (the plane containing the satellite position and velocity vectors) as shown in Figure 2-4. One of the results of Newton's laws shows that *the angular momentum of a satellite's orbit is constant,* which for a vector means constant in both *magnitude* and *direction.* This implies that *an orbit lies in a plane which remains inertially fixed in space.*

Consider, in Figure 2-5, that the vector **h** shown represents the angular momentum of a satellite in orbit around the earth. (We are looking edge-on to the satellite's orbit so its orbital plane looks like just a line across the

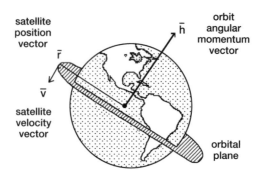

Figure 2-4. The angular momentum vector of an orbit is perpendicular to the orbital plane.

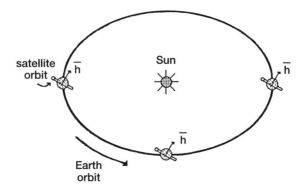

Figure 2-5. Constant angular momentum. This property of an orbit results in an inertially-fixed orbital plane.

earth in the figure.) The figure shows that in half a year, as the earth has moved from one side of the sun to the other, the angular momentum (and thus the plane) of the satellite's orbit has remained pointing toward the same inertial direction in space.

During this time, the earth has also been rotating daily beneath the satellite's fixed orbital plane. This affects the satellite's *ground track* (the path which the satellite appears to etch across the earth's surface), a point which we will discuss shortly.

Total Energy. Another important result of Newton's laws states that *the total energy of an orbiting body is also a constant.* The expression for the *specific total energy* **E** of an orbiting body is:

$$E = \frac{v^2}{2} - \frac{\mu}{r} \qquad (2\text{-}8)$$

The first term on the right-hand side of the equation can be seen to relate to the *kinetic* energy of the orbiting body and the second term represents the body's *potential* energy. For the total energy of an orbiting body to remain constant, the kinetic energy term must decrease as the potential energy term increases. (Since the potential energy term is a negative value in the equation above, this term will increase [become less negative] as the orbital radius **r** increases.) This corresponds to what was stated earlier in Kepler's second law and demonstrated in Figure 2-3.

Orbital Velocities. Another very useful expression for the total energy is:

$$E = \frac{-\mu}{2a} \qquad (2\text{-}9)$$

Relating this to equation 2-8, we can write:

$$\frac{v^2}{2} - \frac{\mu}{r} = \frac{-\mu}{2a}$$

and solving for velocity we get:

$$v = \sqrt{2\mu\left(\frac{1}{r} - \frac{1}{2a}\right)} \qquad (2\text{-}10)$$

Equation 2-10 is a general equation for velocity at any point in an orbit based on its orbital radius **r**. This equation can be simplified for three specific cases of particular interest:

- In a circular orbit we know that **r** = **a** always, so equation 2-10 may be rewritten as:

$$v_{circ} = \sqrt{\frac{\mu}{r_{circ}}} \qquad (2\text{-}11)$$

This simple relationship shows that *circular orbital velocities* v_{circ} *are constant and determined by the circular orbital radius* r_{circ} *only.*

- From equation 2-5, we know that at apoapsis in an elliptical orbit, **r** = r_a = **a(1 + e)**. Using this in equation 2-10, we can write:

$$v_a = \sqrt{\frac{\mu(1-e)}{a(1+e)}} \qquad (2\text{-}12)$$

Since this corresponds to the point of maximum orbital distance, Kepler and Newton's laws tell us that this apogee velocity v_a is the *slowest* during the orbit.

- Similarly, using **r** = r_p = **a(1 − e)** for the orbital radius at periapsis, equation 2-10 becomes:

$$v_p = \sqrt{\frac{\mu(1-e)}{a(1+e)}} \qquad (2\text{-}13)$$

Being associated with the minimum orbital radius, the periapsis velocity v_p is the *fastest* velocity in the orbit.

We now have enough information to analyze established orbits if given just a few of the orbital parameters.

Example Problem:

Due to thrust limitations and the reaches of the atmosphere, the Space Shuttle is limited to operations between about 200 km to 800 km *altitudes*. From this information, determine the orbital parameters associated with an elliptical orbit between these two altitudes.

Solution:

a. Semi-major axis (equation 2-1):
   a = 6,878 km

Note: *altitudes* must be converted to orbital *radii* by adding the radius of the earth, **R_e = 6,378 km** for use in equation 2-1.

b. Orbital period (equation 2-6):
   T = 5,677 sec = 94.6 min

Slightly over an hour and a half between sunrises!

c. Total specific energy (equation 2-9):
   $E = -28.98$ km^2/sec^2

Note: Like eccentricity, knowledge of the total energy can indicate the *type* of orbit as well. All "closed" (elliptical and circular) orbits have *negative* values for total energy. Zero or positive total energies indicate the "open" (parabolic or hyperbolic) orbits introduced earlier.

d. Eccentricity (using equation 2-3 or equation 2-5):
   e = 0.0436

This value (between 0 and 1) indicates an elliptical orbit and agrees with the closed orbit indicated by the negative value obtained for the total energy.

e. Apogee velocity (equation 2-12):
v_a = 7.29 km/sec = 26,235 km/hr

f. Perigee velocity (equation 2-13):
v_p = 7.95 km/sec = 28,628 km/hr

These are appreciable velocities and the velocity difference between apogee and perigee is also not insignificant!

ORBITAL ELEMENTS

To completely describe the shape and orientation of an orbit around the earth, six quantities must be specified as shown in Figure 2-6. During the following discussion keep in mind that, once set, the plane in which an orbit lies remains inertially fixed in space.

As stated earlier, the *eccentricity* **e** and *semi-major axis* **a** define the shape and size, respectively, of the orbit. The *orientation of the orbital plane* is described by two angles: the *inclination* **i** and the *longitude of the ascending node* Ω. The inclination describes the angle that the orbital plane makes with a reference plane (chosen to be the plane of the equator for earth-orbiting satellites). This is the same as the angle between the angular momentum vector of the orbit **h** and a coordinate axis perpendicular to the reference plane (**K** chosen in the direction of the North Pole) as shown in Figure 2-6.

The longitude of the ascending node is an angle that describes the rotation of the orbital plane from a line within the reference plane which points toward an inertially fixed direction (chosen toward the direction of the *vernal equinox,* an essentially inertially fixed direction infinitely far away in space) to the line formed by the intersection of the reference and orbital planes (*line of nodes*) on the side where the orbital motion is from south to north ("ascending" node).

The *argument of perigee* ω describes the *orientation of the elliptically-shaped orbit within the orbital plane.* It is measured as the angle from the line of nodes to the radius of perigee of the orbit.

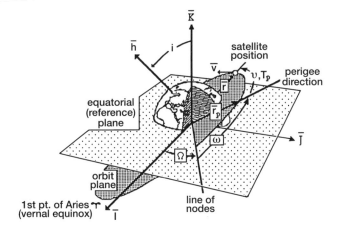

Figure 2-6. Orbital elements. Six independent quantities are required to completely describe an orbit.

Finally, a parameter must be used to describe the *actual position* of the orbiting body at any particular time or location within the orbit. Either an angular measurement, such as the *true anomaly* υ, or a time parameter, such as *time since perigee passage* $\mathbf{T_p}$, may be specified.

ORBITAL PROPERTIES

This section discusses some of the properties of a satellite due to its unique orbital position above the earth.

Field of View

Perhaps the most obvious benefit of being in space is the view one has when looking back at the earth. The amount of the earth in view at any one time is a function of the height above the surface as illustrated in Figure 2-7. In this figure, $\mathbf{R_e}$ represents the mean radius of the earth (6,378 km or 3,444 nm), and $\mathbf{h}$ is the height of a satellite above this mean radius. The point directly beneath the satellite on the surface of the earth is known as the *nadir.*

The angle ϕ is known as the *earth angle* and represents the maximum angular portion of the earth, measured to the tangential horizon, visible from the satellite height. The equation for determining ϕ for any given altitude is:

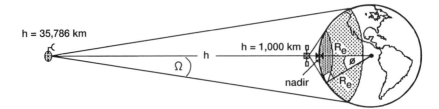

Figure 2-7. Field of view. Properties that a spacecraft has with respect to the surface of the earth change dramatically with altitude.

$$\phi = \cos^{-1}\left(\frac{R_e}{R_e + h}\right) \qquad\qquad (2\text{-}14)$$

The maximum *distance* visible from one tangential horizon to the other is known as the *swath width* **SW** which can be found from:

$$SW = 2R_e\phi \qquad\qquad (2\text{-}15)$$

The *area* on the surface of the earth associated with the swath width is shown shaded in Figure 2-7. This is the *footprint area* of coverage theoretically possible from the satellite position. This area can be found from:

$$A = 2\pi R_e^2 \sin\phi \qquad\qquad (2\text{-}16)$$

Finally, the *angle* subtended (from the perspective of the satellite) between the line from nadir to a line toward the tangential horizon is known as the *angular field of view* (FOV) Ω and is determined from:

$$\Omega = \sin^{-1}\left(\frac{R_e}{R_e + h}\right) \qquad\qquad (2\text{-}17)$$

This angle is important, as it represents the angle that a particular sensor would have to be able to scan or view in order to cover all the potentially visible surface of the earth.

All of these terms have similar but slightly different meanings when used to describe the capabilities of a particular sensor in space, as discussed in Chapter 6. For instance, swath width may be used to describe only the side-to-side distance over the surface of the earth scanned by the sensor, which may be much less than the geometric swath width described here. The same is true of the footprint area viewed by a particular sensor. When discussing the angles through which a sensor may scan, the sensor field of view may be different from the angular field of view described here, and the angles viewed may also be identified as the nadir angle or the look angle. Usually, the context in which these terms are used will clarify the intended meaning.

Ground Track

As mentioned earlier, the *ground track* represents the path of the satellite superimposed onto the surface of the earth. This path is a result of a combination of the orbital motion of the satellite and the movement of the earth due to its rotation. Figure 2-8 shows the ground track of a typical low-earth orbiting satellite through four orbits. The orbit is circular with an altitude of about 1,000 km in an orbital plane inclined about 45° from the equator.

The most notable feature of such a ground track is the apparent movement of the orbital path toward the west. As we learned earlier, the plane in which an orbit lies is inertially fixed in space. The apparent westward movement of the ground track is caused by the eastward rotation of the

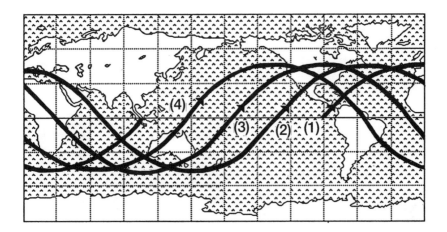

Figure 2-8. Ground track. The path an orbiting satellite traces across the surface of the earth changes due to the earth's rotation.

earth beneath the orbital plane. The distance between the first ascending (south to north) passage through the equator shown in Figure 2-8 (1) and the next ascending passage (2) is determined by the period of the orbit which, as we saw in equation 2-6, is a function of the orbital semi-major axis only.

Another point to notice in Figure 2-8 is that the ground track is constrained between about 45° north and 45° south latitudes. This corresponds directly to the orbital inclination, and constrains the ground track only; the area on the surface of the earth visible to the satellite is a function of the height of the satellite within the orbital plane as mentioned earlier.

Maximum Time in View

The footprint of the satellite's view on the earth moves continuously with the satellite as it travels around the earth. This means that the satellite can only view a particular spot on the surface (or conversely, a ground observer could only see an orbiting satellite) for a limited amount of time. This amount of time depends on the satellite's altitude (which corresponds to the satellite velocity) and orbital inclination, as well as on the latitude and distance from nadir of the ground observer during the satellite pass. For planning purposes, the *maximum* time that a satellite can be in view, associated with tangential horizon to tangential horizon passage directly over the observer's location, can be found from:

$$t_{view} = \frac{2\cos^{-1}\left(\dfrac{R_e}{R_e + h}\right)}{\dfrac{360}{84.49}\left(\dfrac{R_e}{R_e + h}\right)^{3/2} - \dfrac{260}{1436}} \qquad (2\text{-}18)$$

Note that use of this equation requires computation of the arc cosine in degrees. Of course, any oblique pass not directly over the observer's location will result in less time that the satellite will be in view.

Number of Revolutions per Day

The number of times a day that a satellite will completely circle the globe depends on its orbital period, which is a function of the semi-major axis, which can be related to the satellite altitude. For circular orbits (only), a simple expression gives the number of revolutions per day as:

$$\text{\# Rev/day} = 16.997 \left(\frac{R_e}{R_e + h} \right)^{3/2} \tag{2-19}$$

Revisit Time

It may be desirable to have a satellite return over a specific point on the earth's surface periodically for monitoring purposes. It may even be desirable to have the satellite pass over the specified point at the same time of day each visit, or remain in view of a spot indefinitely. Using combinations of the orbital elements and parameters, it is possible to create an orbit to meet many of these requirements; however, there is no simple way of describing this process here. Nonetheless, for some simple cases such as discussed in the following sections, it is possible to see how some of these requirements can be approached.

Revisit times are mainly dependent on orbital period **T** which, as we saw, is a function of the semi-major axis of the orbit. If an orbital period is established which divides evenly into the period of one sidereal earth rotation (T_s = 1 sidereal day = 86,164 sec) such that $T_s/T = n$ where **n** is an integer, then after **n** orbits (neglecting any perturbations) the ground track of the orbit will begin to repeat. If **n** turns out to be unity, then the orbital period exactly matches the rotational period of the earth and the ground track repeats itself each orbit. This is known as a *geosynchronous* orbit.

Example Problem:

The lowest altitude at which a satellite may circle the earth in a repeating orbit (though its lifetime would be limited due to atmospheric drag) is 262 km. Determine the footprint angle, swath width, footprint area, angular field of view, maximum time in view, and the number of revolutions per day. Show that the satellite period is an integral multiple of the sidereal day.

Solution (using equations 2-14 through 2-19 and equation 2-6):

$$\phi = 16.15° = 0.28 \text{ rad}$$
$$SW = 3,595.3 \text{ km (use radians for } \phi \text{ in this equation!)}$$

$A = 71,089,195.3 \text{ km}^2$

$\Omega = 73.85° = 1.29 \text{ rad}$

$t_{view} = 8.59$ min (not a lot of time if you have much to communicate between the satellite and a ground station!)

Rev/day = 16

$T = 5,384.7 \text{ sec} = 89.75$ min

$T_s/T = 16$

USEFUL ORBITS

There are many orbits which have come into common use. This section presents four of the most common.

Low-Earth Orbit

There is no specified cut-off altitude, but as the name implies, low-earth orbits (sometimes designated LEO) represent orbits which come relatively close to the surface of the planet. All the manned spacecraft, with the exception of the Apollo missions to the moon, have been in low-earth orbits, as are many of the earth-observing satellite systems. As the previous sections have implied, low-earth orbits are characterized by short orbital periods, many revolutions per day, high orbital velocities, and limited swath areas on the earth's surface.

Geostationary Orbit

In the last section, an orbit was defined which had an orbital period exactly equal to one sidereal day. Any orbit with this period can be called geosynchronous. A *geostationary* orbit is defined as a *circular, geosynchronous* orbit with an inclination equal to zero (*equatorial*). A satellite placed in such an orbit would appear, to an observer on the surface of the earth, to remain stationary in the sky. The higher altitude in Figure 2-7 shows the approximate position of a geostationary satellite and its associated field of view. Geometrically, a satellite at geostationary altitude can see a footprint about 70 degrees north and south of the equator.

The benefits of a satellite so positioned are obvious. A single sensor placed at a longitude equal to the center of the United States could monitor the weather pattern for the entire country continuously. A communications relay station placed at a longitude in the middle of the Atlantic or

Pacific oceans could allow direct communications between countries on either side or ships upon the seas. Many more uses exist, and the popularity of this operating position is producing a crowded area in space.

Polar Orbits

Strictly defined, polar orbits are orbits with an inclination of 90° which would pass over the earth's poles each orbit. In actual use, many orbits with inclinations near 90° which pass over earth's higher latitudes are also called polar orbits. The benefit of such an orbit, besides being able to view the higher latitudes, comes from the fact that the orbital plane is fixed and the earth rotates continuously beneath this plane. If the orbital period is a nonintegral multiple of the sidereal day, then eventually all areas of the globe will pass beneath the orbiting satellite. Of course, integral multiples of the sidereal day can allow a satellite to pass over the same point on the surface at regular intervals. This type of orbit is used regularly by satellites gathering information about the earth and its environment and resources.

Sun-synchronous Orbits

The sun-synchronous orbit can be considered a special type of polar orbit. Our discussion of orbits so far has ignored the small, but present, perturbing effects that affect the motion of an orbiting satellite. In most cases, these perturbations are accepted or simply compensated for by small station-keeping propulsion systems on board the spacecraft. In the case of a sun-synchronous satellite, though, one of these perturbing forces is used to produce a desired effect.

As depicted in Figure 2-9 (greatly exaggerated), the earth is not a perfect sphere, having more mass at the equator than at the poles. For a satellite in an orbit inclined greater than 0° but less than 90°, this situation imparts a force that attempts to "pull" the orbital plane toward the equator. Due to gyroscopic properties, though, this force results not in a change in orbital inclination, but in a precession of the orbital plane around the equator as indicated in the figure.

The effect of the perturbing force depends on the satellite altitude and the inclination of the orbit. A combination of these two can be arranged to produce a rotation of the orbital plane around the equator equal to about

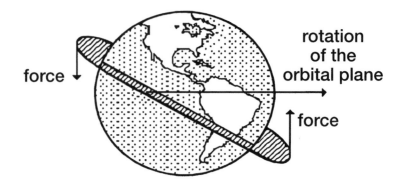

Figure 2-9. Sun-synchronous orbit. The oblateness of the earth causes a rotation of the orbital plane.

one degree per day. This would mean that an orbit established perpendicular to the direction of the sun could be made to maintain this orientation relative to the sun throughout the year. (Refer back to Figure 2-5 which explained the fixed nature of orbital planes.) If a satellite were placed in such an orbit at an altitude producing a period that was an integral multiple of the sidereal day, then that satellite could be made to pass over the same spot on the earth at the same solar time(s) each day, every day of the year. Many remote sensing satellites are placed into such an orbit so as to be able to control the conditions under which the information about an area of interest is gathered throughout the year.

Constellations

As stated earlier, using combinations of the orbital elements and parameters, an orbit can be designed to give a desired coverage of the earth. But there are purposes which require that the entire earth be covered at all times, which is clearly impossible for a single satellite. In these cases, a constellation of satellites is necessary. A constellation is simply a series of two or more satellites purposely placed into mutually supporting orbits. The satellites may be placed within the same orbital plane to effectively decrease the functional orbital period, or satellites may be placed into different orbital planes to compensate for the movement of the observation area due to the rotation of the earth. Figure 2-10 illustrates a constellation employing both these options (three planes of three satellites each).

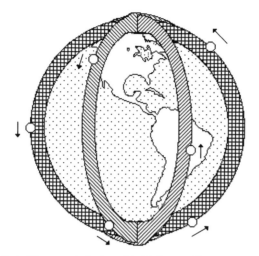

Figure 2-10. Satellite constellation. Multiple satellites can be placed into mutually supporting orbits.

ORBIT ESTABLISHMENT AND ORBITAL MANEUVERS

This section is included to give an orbital mechanics view of orbit establishment and orbital maneuvers. This topic is discussed again in Chapter 3 which deals with the propulsion concepts needed to perform these maneuvers.

Orbit Establishment

It was stated earlier that the total specific energy of an object in orbit was constant and could be determined from equation 2-8, which is repeated here for convenience.

$$E = \frac{v^2}{2} - \frac{\mu}{r}$$

For spacecraft, this energy is usually established by having a launch system carry the payload to a specific altitude $r = R_e + h$ with a specific velocity v before burning out. It must be remembered that the velocity is a vector quantity having both magnitude and direction. If the launch vehicle establishes a velocity magnitude and direction that corresponds exactly with the conditions for the desired orbit at that point, then the desired orbit is achieved. If these conditions are not met, some kind of orbital maneuver may be required.

Orbital Maneuvers

The basic purpose of the launch system is to carry a payload from the surface of the earth into space. However, in most cases, the launch system may not place the payload exactly into the desired orbit or may require more than just the launcher stages to achieve the desired orbit. Also, after establishing an orbit, adjustments may be required so that the orbit will conform with some desired orbital characteristic, such as period or eccentricity, to modify the present orbit to a completely new orbit, or merely to compensate for perturbing forces to maintain the desired orbit.

Once again, as with orbit establishment, it is simply a matter of ensuring that the spacecraft has the proper velocity (both magnitude and direction) at the proper orbital radius at the proper time. This is usually accomplished by adjusting the spacecraft's orbital velocity magnitude or direction until the prescribed conditions are met. This is where the term *delta v* (Δv) comes from when discussing orbital maneuvers.

Orbit adjustments generally fall into two types: in-plane velocity adjustments and plane changes. Addressing all possible combinations of maneuvers between different types of orbits is beyond the intent of this book; however, some basic maneuvers shall be presented to illustrate the general process.

Single-Impulse Maneuvers. Most orbital maneuvers can be considered impulsive (instantaneous) because the actual propulsion system "burn" is relatively short when compared to the typical orbital period. This allows us to assume that the change in velocity during an orbital maneuver occurs at a single point. This also tells us that the initial orbit and final orbit share at least the point at which the maneuver is performed. If the initial and desired orbits share at least one point, then a single-impulse maneuver is all that is required to change from one orbit to the other. The Δv required, which may be an in-plane adjustment, a plane change, or both, can easily be computed using the velocities (magnitude and direction) of each orbit at the common point.

If the initial orbit and the desired orbit do not intersect, it will not be possible to transfer between orbits with a single maneuver. Instead, an intermediate (transfer) orbit that intersects the initial and desired orbits will have to be chosen and multiple Δv's performed at these intersections. However, each of these Δv's may still be treated as separate single-impulse maneuvers.

Hohmann Transfer. A commonly used orbital maneuver between nonintersecting orbits is the Hohmann transfer. The classic Hohmann transfer is characterized by an elliptical transfer orbit between two co-planar circular orbits as shown in Figure 2-11.

The elliptical transfer orbit also lies in the same plane as the other orbits with its perigee at the same altitude as the low orbit and its apogee at the higher orbit altitude. At the intersections (perigee and apogee) of the transfer orbit and the circular orbits, the orbital velocities are parallel and in the same direction. The Δv's are simply the differences in the velocity magnitudes at these points, with no change in direction of the velocity vectors required.

The change in velocity required to enter the transfer orbit from the low circular orbit can be found from the relationships already discussed:

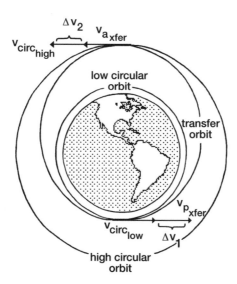

Figure 2-11. Hohmann Transfer. This is a common method for transferring between two co-planar circular orbits.

$$\Delta v_{1_{Hohmann}} = v_{P_{transfer\ orbit}} - v_{circ_{low\ orbit}} \qquad (2\text{-}20)$$

At the transfer orbit apogee, the Δv required to enter the circular higher orbit would be:

$$\Delta v_{2_{Hohmann}} = v_{circ_{high\ orbit}} - v_{a_{transfer\ orbit}} \qquad (2\text{-}21)$$

Simple Plane Changes. A *simple* plane change only changes the inclination of the orbit and does not affect the other orbital characteristics. Plane changes must also be done at the intersection of the initial orbit and the desired orbit, as depicted in Figure 2-12. The Δv required for a simple plane change can be found from:

$$\Delta v_{simple\ plane\ change} = 2v\sin\frac{\Delta i}{2} \qquad (2\text{-}22)$$

where v represents the orbital velocity and Δi is the change in inclination. An important point that should be realized from equation 2-22 is that plane changes performed at low orbital velocities minimize the required Δv. This suggests that plane changes should be performed at high altitudes or at points of lowest velocity. It should also be realized that plane changes in general are expensive in terms of Δv. Equation 2-22 shows that a plane change of 60° requires a Δv equal to the orbital velocity itself!

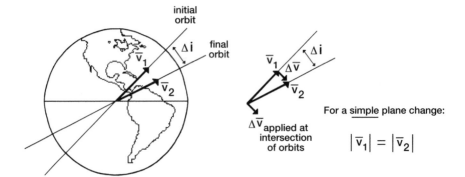

Figure 2-12. Plane change. Plane change maneuvers are performed over the equator where the two orbital planes intersect.

REFERENCES/ADDITIONAL READING

Boorstin, D., *The Discoverers.* New York: Vintage Books, 1985.

Bate, R., Mueller, D., and White, J., *Fundamentals of Astrodynamics.* New York: Dover Publications, Inc., 1971.

Eisele, J., and Nichols, S., *Orbital Mechanics of General-Coverage Satellites.* Naval Research Laboratory Report 7975, Washington, D.C.: U.S. Government Printing Office, 1976.

EXERCISES

1. Due to drag considerations, the lowest altitude at which the Space Shuttle can operate is around 200 km. For a shuttle in a circular orbit at this altitude, determine the following orbital parameters:

   a. semi-major axis
   b. orbital period
   c. total energy
   d. eccentricity
   e. orbital velocity

2. Due to thrust considerations, the highest altitude at which the Space Shuttle can operate is around 800 km. For a shuttle in a circular orbit at this altitude, determine the same orbital parameters as above.

3. Compare each answer obtained in the above problems. Discuss the differences in the answers and the reasons for the differences.

4. On February 20, 1962, John Glenn became the first American to orbit the earth. At apogee, his altitude reached 260.9 km and his perigee altitude was 160.9 km. Determine the following parameters for his orbit:

   a. semi-major axis
   b. orbital period
   c. total energy
   d. eccentricity
   e. perigee velocity
   f. apogee velocity
   g. the maximum time that Glenn would be in communication range (in view) of a ground station at any time during the orbit (this would be at apogee)

5. Determine the following properties for a circular *geosynchronous* orbit:

    a. orbital period
    b. semi-major axis
    c. total energy
    d. eccentricity
    e. orbital velocity
    f. # rev/day

6. Determine the following properties for a *geostationary* orbit:

    a) altitude
    b) footprint angle
    c) swath width
    d) footprint area
    e) field of view
    f) inclination
    g) what about the other orbital elements (Ω, ω, ϕ, etc.)?

CHAPTER 3

Propulsion

When Newton theorized the possibility of placing an object into an orbit around the earth, not only could very few imagine a purpose for such an endeavor, no one could imagine a way to even attempt the feat. While rockets had been a part of man's history ever since the ancient Chinese had used them for dazzling fireworks displays, uses for such systems had been limited to the purpose (primarily military) of hurling munitions from one point to impact another on the surface of the earth. In fact, this was still the main purpose even after technology allowed man to consider Newton's musings possible.

The modern ideas of rocket propulsion actually had their beginnings in the 1880s in the small Russian town of Kaluga, south of Moscow. There, Konstantin Tsiolkovsky worked out the fundamental laws of rocket propulsion and published his work proving the feasibility of achieving orbital velocities via rockets in 1903—the same year of the Wright brothers' first successful powered flight! Tsiolkovsky had earlier accurately described the phenomenon of weightlessness in space (1883), predicted earth satellites (1895), and suggested the use of liquid hydrogen and oxygen as propellants, which three-quarters of a century later were used to send men to the moon.

With no knowledge of the work of the Russian scientist, Robert H. Goddard began studying rocketry shortly before World War I. In 1919, the Massachusetts physics teacher sent a 69-page treatise to the Smithsonian Institution entitled "A Method of Reaching Extreme Altitudes." The "extreme altitude" he was referring to turned out to be none other than the moon, and his treatise earned him a great deal of ridicule from the press at the time. On March 16, 1926, Goddard launched the first liquid-fueled rocket in history which burned for 2.5 seconds and landed a few hundred feet away in his aunt's cabbage patch. The press, once again, had a good time with headlines like "Moon Rocket Misses Target by 238,799½ Miles." Nonetheless, Goddard, backed by grants from the Guggenheim

Foundation arranged by pioneer aviator Charles Lindbergh, continued his work with liquid-fueled rockets (in seclusion from the press), solving many of the practical problems involved with such an endeavor.

The third rocket pioneer of note was Hermann Oberth, a German schoolteacher, whose paper "The Rocket into Interplanetary Space" (1923) gained him the attention of his country and whose research culminated in the development of the V-2 rocket used by Hitler to terrorize England during the second world war.

Also instrumental in the development of the V-2 was Wernher von Braun, who surrendered to American troops in 1945 and went on to become a driving force in U.S. rocket development efforts. Von Braun's first U.S. rocket, the Redstone (which later carried America's first man into space), was ready for launch in 1956 but, for military and political reasons, was not allowed to launch until much later. Nonetheless, von Braun was instrumental in the development of the Jupiter C rocket that placed America's first satellite, *Explorer 1,* into orbit in January 1958 and the development of the Saturn family of rockets which were the mainstay of the Apollo program's goal to reach the moon.

But it was early in the morning of October 4, 1957, that the "Space Age" truly began when the first beeps from the Soviet spacecraft *Sputnik 1* could be heard around the globe. In those Cold War days, America was alarmed by the obvious potential this represented for a power possessing nuclear capabilities. The main U.S. space effort at that time was the Navy's Project Vanguard, which was suddenly pressured to match the Soviet achievement. The attempt proved premature as the rocket blew up on the pad on December 6, 1957, in front of a television audience of millions. Although Vanguard was eventually successful, it was the Army's Explorer program, headed by von Braun, which brought America into the Space Age and began the efforts, fueled by national competition, to create larger and more capable rockets for purposes of launching machine and man into the environment of space.

FIRST PRINCIPLES

Whatever space mission is undertaken, the spacecraft must first be put into an orbit and, secondly, may need to maneuver as well. For this it requires some sort of propulsion system and, in most cases, more than one. While there are a number of advanced types in development, the vast majority of propulsion systems today produce thrust by the expulsion of chemical propellants through nozzles. Processes for imparting energy to the propellant include chemical combustion, nuclear, electric, and others, and the lists of available propellants is also long and varied. Before we can look at these points, the principles of producing thrust via the expulsion of mass must first be understood.

The Rocket Equation

Newton's second law of motion was discussed in the previous chapter. The law states that *an applied force will result in a change in a body's momentum*. Stated mathematically:

$$\overline{F} = \frac{d\overline{\rho}}{dt} = \dot{\overline{\rho}} \tag{3-1}$$

where momentum is defined as the product of mass and velocity: $\overline{\rho} = \mathbf{m}\overline{\mathbf{v}}$. The dot over the momentum term in equation 3-1 indicates the time rate of change of that quantity, and the line over a term indicates that it is a *vector* quantity having both magnitude *and* direction.

In a situation where the mass does not change, equation 3-1 reduces to the familiar form $\overline{F} = \mathbf{m}\overline{\mathbf{a}}$, where $\overline{\mathbf{a}}$ represents the acceleration of the mass. However, as a rocket ejects exhaust gases for propulsion, the mass term is not constant. To determine the change in momentum, and thus the associated force, we must look at the situation at two different instances in time as shown in Figure 3-1.

The figure illustrates a rocket at time **t** having a velocity $\overline{\mathbf{v}}$, and the same rocket at time $\mathbf{t} + \Delta\mathbf{t}$ with a new velocity $\overline{\mathbf{v}} + \Delta\overline{\mathbf{v}}$ having exhausted an amount of propellant $\Delta\mathbf{m}$, which now has its own velocity $\overline{\mathbf{v}}_0$. If we consider a *closed* system consisting of both the rocket and the ejected pro-

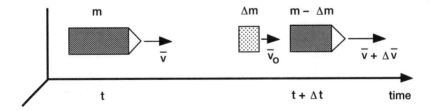

Figure 3-1. Rocket momentum. A rocket trades momentum with its exhausted propellant to produce thrust and increase velocity.

pellants at all times, momentum must be conserved from one time instant to the next. At time **t** the momentum is:

$$\overline{\rho}_{(t)} = m\overline{v}$$

and at time Δ**t:**

$$\overline{\rho}_{(t + \Delta t)} = (m - \Delta m)(\overline{v} + \Delta\overline{v}) + \Delta m\overline{v}_0$$

The change in momentum is:

$$\Delta\overline{\rho} = \overline{\rho}_{(t + \Delta t)} - \overline{\rho}_{(t)}$$
$$= m\overline{v} + m\Delta\overline{v} - \Delta m\overline{v} - \Delta m\Delta v + \Delta m\overline{v}_0 - m\overline{v}$$

which, if we ignore the product of two differential terms, reduces to:

$$\Delta\overline{\rho} = m\Delta\overline{v} + (\overline{v}_0 - \overline{v})\Delta m$$

Dividing through by Δ**t** and taking the limit as Δ**t** approaches zero:

$$\frac{d\overline{\rho}}{dt} = m\frac{d\overline{v}}{dt} + (\overline{v}_0 - \overline{v})\frac{dm}{dt}$$

or, rewriting:

$$m\frac{d\overline{v}}{dt} = \frac{d\overline{\rho}}{dt} + (\overline{v} - \overline{v}_0)\frac{dm}{dt} \qquad (3\text{-}2)$$

Looking at each term of equation 3-2, we see that the term on the left-hand side of the equation represents the instantaneous *acceleration* of the rocket mass **m.**

   The first term on the right-hand side of the equation represents the change of momentum *for the entire system.* Since we are considering a closed system consisting of the rocket and exhaust gases at all times, a change in the overall system momentum can only be caused by forces external to the described system. This term represents *external forces* ($\overline{F}_{ext}$) such as gravity, drag, solar pressure, and others. The vector nomenclature shows that force direction must be understood. Some of these forces will be opposite the direction of motion, such as atmospheric drag, but some may have positive components in the direction of intended motion as can happen with solar pressure. Another external force, due to a difference between the propellant exhaust pressure and the external ambient pressure, may also exist as we will soon see.

The remaining term represents the change in momentum of the exhaust gases. This exchange of momentum is the main contributor to the acceleration of the rocket and this last quantity is known as the *thrust* term ($\overline{T}$). Note that the (vector) quantity $\overline{v}_0 - \overline{v}$ would be negative since the exhaust gasses are ejected opposite the direction of the original velocity. Rewritten as $\overline{v} - \overline{v}_0$, this quantity describes the velocity of the exhaust gasses relative to the rocket itself. We can define this term as the *exhaust velocity* $\overline{v}_e$.

With the definition of terms given above, equation 3-1 can be rewritten as:

$$m\frac{d\overline{v}}{dt} = \overline{F}_{ext} + \overline{T} \qquad (3\text{-}3)$$

which, in this simple form, is known as the *rocket equation.*

Rocket Thrust

Looking more closely at the results of the last section, it can be seen that the thrust term of the rocket equation is proportional to both the propellant exhaust velocity ($\overline{v} - \overline{v}_0 = \overline{v}_e$) and the *mass flow rate of propellant* ($dm/dt = \dot{m}$). We can rewrite the thrust term to better show this dependency:

$$\overline{T} = (\overline{v} - \overline{v}_0)\frac{dm}{dt} = \overline{v}_e\dot{m} \qquad (3\text{-}4)$$

To increase the thrust of a rocket then, one could try to increase either the exhaust velocity of the propellant or the mass flow rate of propellant through the rocket. To see how these quantities can be changed we must consider the characteristics of a typical rocket system such as that shown in Figure 3-2.

Mass Flow Rate. A rocket differs from a jet engine in that a rocket must carry its own oxidizer as well as fuel supply, although there are some systems which simply use a single (mono) propellant. In many propulsion systems, liquid propellants are used and are delivered to the combustion chamber by mechanical pumps as depicted in the figure. The pumps con-

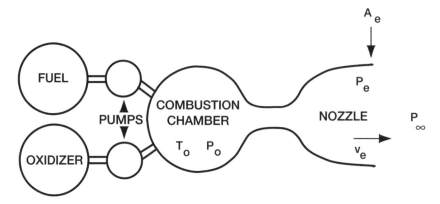

Figure 3-2. Sketch of a typical rocket. Propellants are mixed in the combustion chamber and accelerated through the nozzle to produce thrust.

tribute directly to the rocket thrust by controlling the mass flow rate of the propellant. Lower-thrust systems may use a pressurized bladder or gravity-feed to deliver propellants to the combustion chamber. In solid rocket motors, the solid fuel and oxidizer materials are premixed and loaded into the motor casing which also serves as the combustion chamber when the fuels are ignited. Mass flow rate in solids is established by controlling the burning area and, thus, the combustion rate of the fuel.

There are some obvious limitations to increasing thrust by increasing the mass flow rate. For instance, you could burn all the fuel at once, assuming you could design a combustion chamber and nozzle to handle the amount of propellant and exhaust involved. However, if you didn't just blow up, the instantaneous thrust would probably produce an unacceptable acceleration in the view of the structural designer or payload/astronaut. More practically, the size and structural capabilities of the pumps in liquid-fueled systems limit the mass flow rates achievable. Pumps also contribute to thrust by affecting the exhaust velocity, as discussed next.

Exhaust Velocity. The propellants are combined, and if necessary ignited, within the combustion chamber where they create a high temperature (T_0), high pressure (p_0) mixture. The velocity of the propellants in the combustion chamber is relatively slow and can be considered essentially zero, but the mixture expands to supersonic speeds through the convergent-divergent nozzle, leaving with an exit velocity v_e at an exhaust pressure p_e. The nozzle exit area is represented by A_e in Figure 3-2.

Aerodynamic and thermodynamic relationships relate the conditions between the combustion chamber and the exit area of the expansion nozzle. Though the derivation is beyond the scope of this introduction, the exhaust velocity can be found to be:

$$v_e = \sqrt{\frac{2\gamma RT_0}{\gamma - 1}\left(1 - \left[\frac{p_e}{p_0}\right]^{\frac{\gamma - 1}{\gamma}}\right)} \qquad (3-5)$$

An examination of this relationship tells us what is necessary to have a high propellant exhaust velocity, and thus, higher thrust.

The term γ is the *ratio of the specific heats* of the propellant mixture:

$$\gamma = \frac{c_p}{c_v}$$

where: c_p = specific heat (under constant pressure)
c_v = specific heat (under constant volume)

The particular value for γ depends on the properties of the particular propellant, but its value is usually between 1.2 to 1.4 and for our purposes can be simply considered a constant in equation 3-5.

The term **R** represents the *specific gas constant* for the particular propellant used. This term is actually the Universal Gas Constant (R = 8.314 joule/mole °K) divided by the *molecular weight* of the propellant (M kg/mole). From equation 3-5 we see that a propellant with a lower molecular weight will contribute to a higher exhaust velocity. This shows the benefit of using hydrogen as a propellant, as first realized by Tsiolkovsky, over heavier fuels like hydrocarbons.

A higher combustion temperature T_0 increases the exhaust velocity directly. This temperature is known as the *adiabatic flame temperature* and is primarily a function of the propellants used and their combustion properties. Practical limitations in combustion temperatures exist due to the structural properties of the combustion chamber and rocket nozzles. Liquid oxygen (LOX) and liquid hydrogen are kept at very low temperatures, and many liquid-fueled systems circulate these propellants through tubes surrounding the combustion chambers and nozzles to keep their temperatures within acceptable limits.

Inside the brackets of equation 3-5 is a ratio of exhaust pressure to combustion chamber pressure. Minimizing this ratio contributes to higher exhaust velocities which calls for as large a combustion chamber pressure as possible. This pressure is produced by the combustion of the propellants; however, as was mentioned earlier, the pumps in a liquid-fueled system affect the combustion chamber pressure as well as the mass flow rate in contributing to rocket thrust. Again, limitations to achievable combustion chamber pressures are structural in nature.

Specific Impulse. We can define the term *specific impulse* (I_{sp}) as the thrust produced per time rate of change of *weight*. Remembering that thrust can be related to exhaust velocity, we can write:

$$I_{sp} = \frac{T}{\dot{w}} = \frac{v_e \dot{m}}{\dot{m}g} = \frac{v_e}{g} \text{ (sec)} \tag{3-6}$$

The specific impulse changes for different propellants and different rocket designs and their computation is quite complicated. However, specific impulse is the most commonly used criteria for comparing rocket systems since the resulting units are merely seconds. Table 3-1 shows some of the average properties of three propellant combinations.

Table 3-1
Fuel/Oxidizer Properties

Fuel/oxidizer	T_o (°K)	M (g/mole)	I_{sp} (sec)
Kerosene/oxygen	3,144	22	240
Hydrogen/oxygen	3,517	16	360
Hydrogen/fluorine	4,756	10	390

The Russians use the kerosene/oxygen combination of fuels exclusively in their launch boosters. These propellants were also used on the first stage of the Saturn V rocket used in the Apollo missions to the moon, but the United States regularly uses liquid hydrogen and oxygen for propellants in a majority of their launch vehicles including the main engines of the Space Shuttle. The difference in the use of these fuels is mostly due to their ease of handling; liquid kerosene (also known as RP-1) is much easier to store and transport than liquid hydrogen. This is particularly demonstrated by the hydrogen/fluorine combination which has been used in some special situations to take advantage of the higher specific impulse, but is not used regularly due to the highly toxic nature of fluorine.

Example Problem:

The space shuttle main engine (SSME) uses a liquid hydrogen/oxygen propellant combination. Mass flow rate to each engine is 466.6 kg/sec and the combustion chamber pressure is 20.5×10^6 N/m^2. Determine the standard sea-level exhaust velocity, thrust, and specific impulse for a single SSME. (Assume $p_e = p_{s.l.} = 1.01 \times 10^5$ N/m^2; $\gamma = 1.2$; and $g = 9.81$ m/s^2)

Solution:

Using the values for **To, R,** and **M** (converted to kg/mole) for the hydrogen/oxygen fuel combination given in Table 3-1:

$v_e = 3{,}589.4$ m/sec
$I_{sp} = 365.9$ sec
$T = 1.68 \times 10^6$ N $= 376{,}835$ lbs thrust

For comparison, a fully loaded space shuttle (including fuel tank and solid rocket motors) weighs over 4 million pounds, requiring three SSMEs and two solid rocket motors that deliver over one million pounds thrust each just to get off the launch pad.

Nozzle Design

If we look at a closed *volume* approach to the thrust equation, the situation would look like that shown in Figure 3-3. In this case our control *volume* merely envelopes the rocket and we must consider forces which interact with, and through, this boundary. Summing the forces over the entire volume we find:

$$\Sigma F = \dot{m}v_e + (p_e - p_\infty)A_e \tag{3-7}$$

The first term on the right-hand side of equation 3-7 represents the force due to the propellants passing through the control volume at velocity v_e. This should be recognized as the same as the thrust term defined earlier.

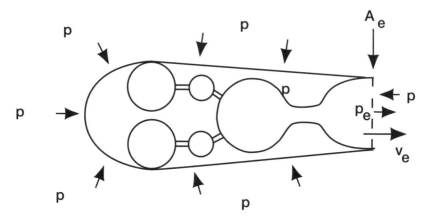

Figure 3-3. Closed volume forces. An unbalanced force may exist on a rocket due to differences between the exhaust pressure and ambient pressure.

The second term shows that the rocket may also feel a force on it due to a difference between the exhaust pressure and the ambient pressure in which the rocket is operating. This term is called the *pressure thrust*. The propellant exhaust pressure (like the exhaust velocity) is due to the aerodynamic and thermodynamic expansion of the propellant through the converging-diverging nozzle, the design of which affects performance depending upon the particular operating environment. Three situations must be considered, as depicted in Figure 3-4.

Overexpansion. In this case, p_e is less than p_∞, indicating that the exhaust gases have been *overexpanded*. This is undesirable because it produces a net pressure force opposite that of the rocket thrust (the p_∞ in front of the rocket is not offset by the p_e at the exhaust area).

Underexpansion. If the exhaust nozzle produces a situation where p_e is greater than p_∞, the propellant gases will have been *underexpanded*. This case looks, at first, to be desirable because the additional pressure force in the same direction as the rocket thrust should help push the rocket forward. However, we may recall that the thermodynamic equation for the exhaust velocity (eq. 3-4) includes a term with the ratio of exhaust pressure over combustion chamber pressure. We wished to minimize this term, which is why a high combustion pressure is desirable, but a high exhaust pressure increases this term. In fact, the loss of thrust due to a

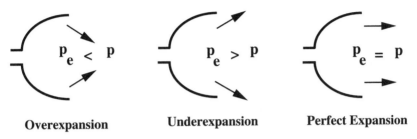

Overexpansion   Underexpansion   Perfect Expansion

Figure 3-4. Nozzle exhaust expansion. It is desirable to minimize the difference between exhaust pressure and ambient pressure.

higher exhaust pressure in this equation far exceeds the gain in thrust due to the pressure thrust term.

Perfect Expansion. In this case, the exhaust nozzle expands the propellant gases such that p_e is exactly equal to p_∞ and the pressure thrust term equals zero. By reason of default of the other cases, this is the best situation when considering the pressure thrust term and nozzle design.

The practicalities of nozzle design mean it is not always possible to maintain a perfect expansion condition throughout the operating range of the rocket. For instance, launch vehicles are used to carry payloads from the surface of the earth, through the atmosphere, into the near vacuum of space. Although variable exhaust area nozzles have been designed to continually match exhaust pressures with the ambient conditions during transit through the atmosphere, these systems are too complex to be practical and the thrust benefit is not worth the expense. In most cases, a median operating pressure is chosen and the nozzle designed to expand the propellants to this pressure, accepting the small changes in thrust produced at other pressures. For systems that operate only in space, the situation is simpler and the nozzle is designed to expand the exhaust to as low a pressure as possible. This results in large, long rocket nozzles as exemplified by the Apollo service module nozzle depicted in Figure 3-5.

ORBIT ESTABLISHMENT AND ORBITAL MANEUVERS

As was mentioned at the end of Chapter 2, a spacecraft's propulsion requirements may include delivery to space, maneuvering into position, and maintenance of the spacecraft position and orientation. While position-keeping systems are usually designed as an integral part of the space-

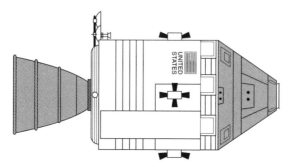

Figure 3-5. Apollo Service Module propulsion systems. The large aft nozzle was used for orbital maneuvers and the smaller nozzles around the body were used for attitude control.

craft, rockets that deliver a spacecraft to its operating position are more likely to be existing systems which are chosen for specific performance characteristics and to which the spacecraft must be designed for compatibility. Specific systems that are available to meet these requirements will be presented in Chapter 8. The following sections describe the propulsion aspects that must be taken into account when discussing these systems.

Orbit Establishment

Several factors are involved in reaching the desired orbit, including establishing the correct burn-out conditions, staging, launch timing, launch pad location, and launch direction. Figure 3-6 depicts a typical launch situation from the U.S. launch site at the Kennedy Space Center (KSC) in Florida.

Burn-out Conditions. The launch vehicle both accelerates and raises the payload from the surface of the earth along a predetermined launch path (trajectory). The launch path must intersect the desired orbit, and the intersection must occur such that the rocket velocity (vector) and altitude correspond with those for the desired orbit *at that point.* Since the rockets cease burning fuel at this point, the velocity and altitude achieved are known as the launch vehicle *burn-out* conditions. If the burn-out conditions do not match the orbit characteristics at that point, within some allowable tolerance, adjustments using supplemental propulsion systems may be required.

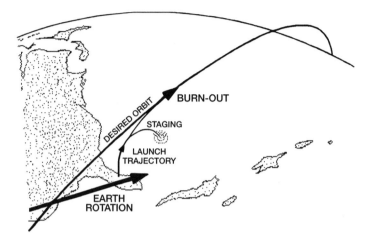

Figure 3-6. Orbit insertion. The launch vehicle must intersect the desired orbit at the proper altitude and with the proper velocity (magnitude and direction).

Staging. Multistage rockets are used to more efficiently (and practically) accelerate payloads to the high orbital velocities required. The separate stages are designed to optimize the burn-out velocity of each stage with respect to the overall mass the stage is accelerating. Upon burn-out, the empty stage is discarded and the next stage, optimized with respect to the remaining mass, takes over. The process is repeated until the desired velocity is achieved.

A relatively simple relationship can be used to get an approximation of the number of stages required to deliver a payload to a particular orbit using a propulsion system of known characteristics:

$$MR = \frac{m_o}{m_{bo}} = e^{\left(\frac{v_r}{n v_e}\right)} \qquad (3\text{-}8)$$

In equation 3-8, **MR** represents the *mass ratio* or ratio of the original mass (m_o) to the mass at burn-out (m_{bo}) for each stage. Use of equation 3-8 assumes that this ratio is constant for all stages of the rocket. The velocity required to establish the orbit is represented by v_r. This number will be higher than the orbital velocity of the desired orbit because it takes into account the excess thrust needed to operate through the atmosphere, elevate the payload to the orbital altitude, and redirect thrust during ascent to steer the rocket along the desired trajectory. The rocket exhaust velocity, also assumed constant for all stages, is given by v_e. Finally, **n** represents the number of stages of the rocket.

Example Problem:

Although orbital velocity for a circular orbit at 300 km altitude is 7.73 km/sec, a typical velocity that a launch vehicle may be required to deliver to reach this orbit is 9.5 km/sec. Assuming use of space shuttle main engines for the propulsion system, determine the mass ratio for a single-stage-to-orbit (SSTO) rocket.

Solution:

Using the exhaust velocity found earlier for the SSME and remembering to use compatible units:

MR = 14.11

This indicates that the single stage would have to have about 14 times the mass of the payload in order to reach the given orbital velocity. This is actually a somewhat respectable mass ratio and illustrates the effectiveness of the SSME design. However, it must be remembered that equation 3-8 gives only a crude approximation due to the assumptions mentioned earlier. Actual stage optimization is a complicated, iterative process that takes into account the different thrust and mass ratio characteristics for each stage.

Launch Timing (Windows). In the preceding chapter we saw that the plane of an orbit is fixed inertially in space while the earth rotates beneath this plane. If it is desired to place the spacecraft into an orbital plane with a particular inertial orientation, the launch will have to be timed so as to occur just as the launch site rotates beneath the desired orbital plane, as depicted in Figure 3-6 for a launch from the Kennedy Space Center.

In most instances, a plus or minus time period around the optimum time of launch is specified and is known as the *launch window*. If launch does not occur during this time period, the launch will have to be delayed until the earth rotates the launch site to another intersection with the desired orbital plane. For some launches, such as interplanetary missions, launch opportunities may be infrequent and launch windows may be as small as only a few minutes. Countdowns for launches are based on completion of all prelaunch activities before entering the launch window.

Launch Pad Location and Launch Direction (Azimuth). The *direction* in which the launch vehicle is fired from the launch pad is known as the *launch azimuth*. Azimuths start at 0° for a northerly launch and increase in a clockwise direction around the compass, as depicted in Figure 3-7.

Launch pad location and launch azimuth are important, as they establish the *inclination* of the resulting orbit. This concept is most easily understood by looking at hypothetical launches from three different launch sites. Refer to Figure 3-7 and remember that the plane of the orbit cuts through the launch site and the center of the earth and contains the velocity vector of the spacecraft. Also remember that orbital inclinations are only described between 0° and 180° (refer back to Figure 2-6 in Chapter 2). Finally, the component of velocity imparted to our launch vehicle due to the rotation of the earth is initially ignored in this discussion but shall be reintroduced later.

Equatorial Launch Site. Launch sites located on the equator (0° latitude) can attain any inclination orbit directly by simply launching in that direc-

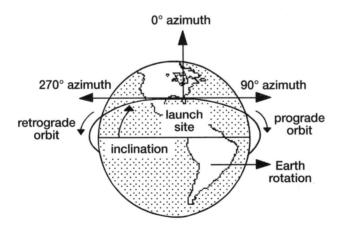

Figure 3-7. Launch azimuth versus orbit inclination. Launch site latitude and launch direction (azimuth) determine orbit inclination.

tion from the pad. For example, a northerly launch (0° azimuth) would send its payload into an orbit passing over the poles continuously (ignoring the component of velocity imparted to our launch vehicle due to the rotation of the earth). This is the polar orbit described in the previous chapter with a 90° inclination. Note that a southerly launch (180° azimuth) also produces a polar orbit. In fact, we could achieve exactly the same (inertially-fixed) orbit as a northerly launch if we simply waited 12 hours for the launch site to rotate to the other side of the earth, and launched south!

An easterly launch (90° azimuth) from this launch site would establish an orbit of 0° inclination—an equatorial orbit—circling the earth over the equator constantly. An orbit with a component of its velocity (vector) in the same direction as the (easterly) rotation of the earth is known as a *prograde* orbit. Launching to the west (270° azimuth) would also establish an equatorial orbit, but the velocity would be opposite that of the rotation of the earth. An orbit with a component of its velocity vector opposite that of the rotation of the earth is called a *retrograde* orbit. The inclination of our retrograde equatorial orbit would be 180°, due to the fact that the angular momentum vector would point opposite that of the prograde equatorial orbit.

Launching from our equatorial launch pad at a 45° azimuth results in an orbital inclination of 45°. Note that launching at 135° azimuth also produces a 45° inclination orbit. This is similar to establishing a polar orbit by launching north, or waiting 12 hours and launching south to enter the same inertially-fixed orbit.

As you may now deduce, from our equatorial launch site we can enter any inclination orbit directly by simply launching in a compatible azimuth direction. You may also have noticed that launch azimuths between 0° and 180° result in prograde orbits with inclinations between 0° and 90°. Launch azimuths between 180° and 360° produce retrograde orbits characterized by inclinations between 90° and 180°.

Polar Launch Site. Next consider a launch site located on the North Pole (90° north latitude). The only direction toward which to launch is south (180° azimuth). This means that the *only* orbit that can be entered into directly from this launch site is a polar orbit of 90° inclination. Notice how the inclination corresponds to the launch site latitude.

Mid-latitude Launch Site. Now consider a launch site located at mid-latitudes, say 28.5° north. If we launch directly north (or directly south) from this point we would still establish a polar orbit of 90° inclination (again ignoring the component of velocity imparted to our launch vehicle due to the rotation of the earth). However, if we launch directly east (90° azimuth), we find that the resulting orbit has an inclination of 28.5° corresponding to the launch site latitude (refer to Figure 3-7). Launching a little northerly or even a little southerly (greater or less than 90° launch azimuth) from this location results in an orbital inclination greater than 28.5°, meaning that 28.5° is the *minimum* inclination orbit into which we can launch directly from this site.

The conclusion one should draw from this discussion is that *the inclination into which a spacecraft can be launched directly is limited by the geographical location (latitude) of the launch site.* If an orbit of lower inclination than the launch site latitude is desired, then an orbital adjustment (plane change) must be performed requiring additional fuel and/or propulsion systems, which generally means increased weight and cost.

The 28.5° latitude example used in the above discussion corresponds to the latitude of the Kennedy Space Center. Figure 3-8 shows the relation-

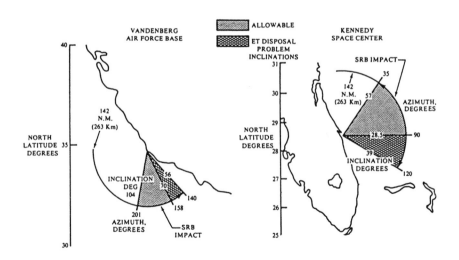

Figure 3-8. United States launch sites. This figure shows the relationships between launch azimuths and orbital inclinations for U.S. launch sites.

ship between launch azimuth and resulting orbital inclinations from both the Kennedy Space Center and Vandenberg Air Force Base in southern California, the two major U.S. launch sites.

Though theoretically unlimited, the actual launch azimuths used by the United States are restricted to the directions that allow jettison (and in the case of the space shuttle, recovery) of the initial stages of the launch vehicle over the ocean. Vandenberg is used primarily to launch spacecraft into polar or the slightly retrograde sun-synchronous orbits, and, for the safety reasons just stated, launch azimuths are always to the south.

Earth Rotation. In the previous discussions we ignored the rotation of the earth for simplicity in relating launch azimuths to orbital inclinations. However, the earth rotates at an appreciable rate which contributes a significant initial velocity to the launch vehicle. This initial velocity assists launches to the east, with the greatest assistance realized for 90° azimuth launches. This easterly component of velocity must be compensated for in order to launch into a truly polar orbit, and must be overcome in order to launch into a retrograde orbit. Launching in any azimuth other than 90° decreases the amount of payload a particular launch system can deliver to orbit compared to a direct easterly launch. This explains the popularity of 28.5° inclination orbits for U.S. space missions.

Orbit Adjustment

It was noted earlier that if the launch vehicle does not establish the desired orbital position and velocity precisely, or if the initial orbit is not the spacecraft's operating orbit, then an orbit adjustment will be required. We differentiate orbital adjustments from station keeping or attitude adjustments by the relatively larger Δv requirements of orbital adjustments, as quantified in Chapter 2. Missions requiring orbital adjustments may incorporate completely separate propulsion systems to accomplish these maneuvers, while station keeping and attitude adjustment (if performed by a propulsion system) may be accomplished by a system integrated into the design of the spacecraft.

Addressing all possible types of orbital adjustments and their associated propulsion requirements is beyond the intent of this book. However, an excellent and commonly performed example exists which can be used to describe both the magnitude of adjustment required and the typical types of systems used to perform these maneuvers.

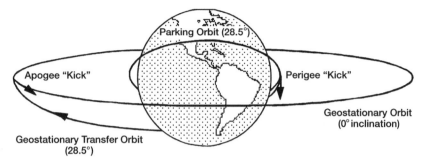

Figure 3-9. Geostationary orbital transfer. The plane-change maneuver is performed out at the apogee of the transfer orbit to minimize the Δv required.

Geostationary Transfer. Figure 3-9 illustrates the transfer of a satellite from a typical low-earth "parking" orbit of 28.5° inclination to a geostationary orbit.

The first maneuver establishes the spacecraft on a *transfer orbit* which is an ellipse that lies in the same plane as the parking orbit with its perigee at the same altitude as the parking orbit and its apogee at geostationary altitude. This should be recognized as a Hohmann transfer orbit described in Chapter 2. Note that in order to intersect the final desired equatorial geostationary orbit, the apogee (and, thus, the perigee) of the 28.5° inclined transfer orbit must be located over the equator. As the satellite approaches this point over the equator, a *perigee kick motor* (PKM) is fired to inject the spacecraft into the transfer orbit from the parking orbit. The change in velocity required at this point can be found from the relationships given in Chapter 2.

At the transfer orbit apogee, two changes to the orbit must occur in order to establish a geostationary orbit. First, the elliptical transfer orbit must be circularized at the geostationary altitude. If this is done as a separate maneuver within the same plane as the transfer orbit, the Δv required would be the same as for a Hohmann transfer. However, to be geostationary, the orbit must also be equatorial, so the plane of our spacecraft orbit must be changed from 28.5° to 0°. The Δv relationship for a simple plane, given earlier in Chapter 2, is repeated here to illustrate an important point:

$$\Delta v_{\text{simple plane change}} = 2v \sin \frac{\Delta i}{2} \qquad (2\text{-}22)$$

This relationship reminds us that plane changes performed at low orbital velocities minimize the required Δv. That is why, in geostationary transfers, the plane change is done out at geostationary altitude and not at the parking orbit altitude. In actual geostationary transfers, an *apogee kick motor* (AKM) is used to perform both the circularization of the transfer orbit and the plane change simultaneously, which results in a savings in Δv required over performing separate maneuvers.

PKMs and AKMs may be either liquid or solid fueled. Because only a single burn of known Δv is required at the "kick" points of a geostationary transfer, solid motors provide a simple way of performing this function and are commonly used. However, if higher performance or a restart capability is required, a liquid-fueled motor may be more suitable. Different PKMs and AKMs as well as currently available orbital maneuvering systems and launch vehicles are presented in Chapter 8.

Before leaving this subject, it is worth noting that the former Soviet Union's major launch site, the Baikonur Cosmodrome in Khazakstaan near the city of Tyuratam southeast of Moscow, is at 45.6° N latitude, as shown in Figure 3-10. Due to range safety reasons and the proximity of China to the east, allowable launch azimuths result in a minimum orbital inclination of 51° from Tyuratam. It should come as no surprise that the Russian *Mir* space station is in a 51° inclination orbit. Also, as with the Apollo-Soyuz mission, future joint U.S.-Russian missions, including the international space station, will probably be conducted in 51° inclination orbits, as the penalty paid by the United States to launch into a higher than 28.5° inclination orbit is far less than that which would be required by the Russians to change the plane of their orbit from 51° to 28.5°. As a last note, the French Ariane launch vehicle has become a major launcher of satellites into geostationary orbit. Much of this success is due to the fact that the Ariane is launched from the Guiana Space Center near Kourou, located at 5.2° N latitude on the northeast coast of South America. Since only a small plane change is required to establish a geostationary orbit from this site, more of the payload launched can go into spacecraft weight and less to propulsion systems making the Ariane attractive to many commercial geostationary satellite providers. Figure 3-10 shows this launch site along with the locations and latitudes of some of the world's other current major orbital launch sites.

Station-keeping and Attitude Control

In Chapter 2 we ignored the many sources of perturbations that constantly act on the orbit and orientation of a spacecraft. If it is important

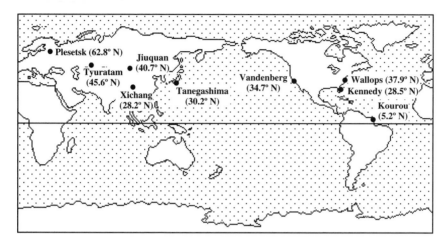

Figure 3-10. Launch sites. Launch site latitude determines the minimum inclination that a satellite may be launched directly into.

that a spacecraft maintain a specific orbit and/or attitude with respect to the earth or stars, then additional propulsion systems may be required. Compared to launch and orbital transfer vehicles, station-keeping and attitude control systems are made up of much smaller thrusters which measure their forces in pounds (or Newtons) or fractions thereof. An orbital transfer vehicle may be capable of delivering a Δv of several kilometers per second in order to make major changes to a spacecraft orbit, but a station-keeping propulsion system is designed to make fine adjustments to an orbit, delivering Δv's measured in meters per second. Attitude control thrusters are even smaller, designed to simply rotate the spacecraft body and not affect the orbit significantly.

As mentioned earlier, these systems are usually custom-designed as an integral part of a spacecraft, and the thrusters must be positioned about the spacecraft body at precise locations to provide the correct torques and magnitudes. If you refer back to Figure 3-5 you will notice the groups of small thrusters placed around the body of the Apollo service module.

Associated with these thrusters, propellant tanks must be included to store the required amount of fuel for the mission lifetime. In many cases, the amount of onboard fuel becomes the limiting factor in useful spacecraft lifetime, and more than one perfectly-operating spacecraft has become useless due to fuel depletion. The propellant used must be capable of on-orbit storage. If a liquid propellant is used, the spacecraft may have to maintain the tanks and lines within acceptable temperatures so the fuel does not freeze and burst its containers. The loss of the $1 billion *Mars Observer* as it was approaching its destination in August 1993 was attributed to burst propulsion system pipes.

Also, since the exhaust gases may remain in the vicinity of the space-craft, the propellant must not become a source of contamination to the spacecraft systems or the operation of the payload. Pressurized gases such as nitrogen, which are less of a source of contamination than many other propellants, are often used for attitude control; however, their performance (specific impulse) is less than that of liquid propellants and may not be acceptable for higher thrust purposes. A commonly used liquid mono-propellant for station-keeping and attitude control is hydrazine (N_2H_4), of which there are several variations (monomethylhydrazine [MMH] and unsymmetrical dimethylhydrazine [UDMH] are common examples). Hydrazine has many desirable characteristics in terms of storage and contamination, and delivers a very acceptable specific impulse of over 200 seconds.

REFERENCES/ADDITIONAL READING

Anderson, J., *Introduction to Flight,* 2nd ed. New York: McGraw-Hill, Inc., 1985.

Sutton, G., *Rocket Propulsion Elements,* 5th ed. New York: John Wiley & Sons, Inc., 1986.

C. Cochran, D. Gorman, and J. Dumoulin (Eds.), *Space Handbook.* Alabama: Air University Press, 1985.

EXERCISES

1. Determine the theoretical exhaust velocity for the liquid hydrogen/liquid oxygen fuel/oxidizer combination given in the reading. Consider the ratio of exhaust pressure to combustion chamber pressure (p_e/p_o) and the ratio of specific heats (γ) to be 0.001 and 1.4, respectively.

2. Recompute the above exhaust velocity with an increase and then a decrease in the pressure ratio of a factor of ten (i.e., $p_e/p_o = 0.01$ and $p_e/p_o = 0.0001$).

3. Compare the answer obtained in Exercise 1 to the exhaust velocities obtained for the other fuel/oxidizer combinations given in the reading. Use the same values for p_e/p_0 and γ given in Exercise 1 above.

4. Compute the specific impulses for the three fuel/oxidizer combinations (based on the answers from Exercises 1 and 3 above) and compare them to those given in the reading. What might be the reason for any differences?

5. The Saturn V F-1 first-stage booster rocket used a kerosene/oxygen propellant combination. Mass flow rate to each engine (there were five in the first stage) was 5,736 lb/sec and the combustion chamber pressure was 1,122 psi. Determine the standard sea level exhaust velocity, thrust, and specific impulse for one of the engines (assume $\gamma = 1.2$). What was the total sea level thrust of the Saturn V? Give your answers in both English and metric units.

6. It is desired to launch a multistage rocket with a ratio of original mass to burn-out mass for each stage of 2.0. For a nozzle design which maintains the exhaust velocity at 3,000 m/sec, determine how many stages would be necessary to place a payload into a circular low-earth orbit at an altitude of 200 km, with a trajectory resulting in $v_r/v_{circ} = 1.1$.

7. Compute the three separate Δv's required for the geostationary transfer described in the reading. Assume an initial circular parking orbit of 450 km altitude at 28.5° inclination.

CHAPTER 4

Spacecraft Environment

Spacecraft operate in an environment totally different from that experienced by systems on earth. In some cases, this environment is relatively benign compared to what earth-based systems must withstand (hurricanes, for example), but in many ways the space environment is harsh on the equipment (and people) sent out to perform there. The space systems engineer must understand this environment and design spacecraft to operate within it. In addition, the spacecraft must be capable of performing its mission *through* this environment. The effects of the environment on typical spacecraft missions will be discussed in subsequent chapters.

THE SUN

The sun is the major environmental influence on the earth and the space around it. This section presents a brief description of the sun and discusses the radiations it emits into space. Following sections describe the properties of these radiations in the vicinity of earth and the environment that results from the interaction of the two. Finally, the effects of this environment on the operation of systems in space is discussed.

Cosmology

Cosmology is the study of the evolution of the universe and includes the theory of how our sun came into being. It is believed that 12 to 18 billion years ago the universe had a violent beginning in an expanding fireball known as the "Big Bang." The Milky Way galaxy represents one of the countless lumps of the products of that explosion, made up mainly of hydrogen and helium. Our solar system began as a cloud of interstellar gas (also mostly hydrogen) which probably contained the debris of the destruction of previous generations of stars along with the primordial elements of the creation of the universe. It is theorized that some sort of shock wave, perhaps from a nearby stellar explosion (the fate of all stars) or a galactic wave front associated with the movement of the spiral arms of our galaxy, disturbed the cloud of gas, causing it to condense in places. These areas of increased density and gravitational attraction pulled in nearby molecules, and the gas cloud began to spiral inward forming a disk-like shape.

Some local clumps of material formed away from the main core and, through *accretion* (collision and fusing) with other clumps, eventually coalesced into what are now the planets and moons. The ring of asteroids around the sun between the orbits of Mars and Jupiter are thought to be a region where this process did not continue long enough to form another planet. The planets all basically lie in the plane of the sun's equator, known as the *ecliptic,* and travel in the same direction around the sun.

As the mass at the center of this activity grew larger, pressure and temperature began to rise as the elements packed in tighter and tighter until conditions reached a critical point and the core burst into spontaneous nuclear fusion. Many more processes occurred in the 4.6 billion years or so it has taken the solar system to stabilize into what we see today, but at that moment a new star was born.

Our sun is a typical star in a typical galaxy containing an estimated 200 billion stars in a universe of perhaps more than a billion galaxies. The sun is located about 30,000 *light-years* (1 l.y. = the distance light travels in one solar year $\approx 9.5 \times 10^{12}$ km) from the center of our spiral galaxy which is about 100,000 l.y. across. The sun has an orbital period of about 200 million years and an orbital velocity of about 250 km/sec around the Milky Way.

Structure of the Sun

The sun has a mass ($\mathbf{M_s}$) of 2×10^{30} kg composed of about 78% hydrogen, 20% helium, and 2% heavier elements by weight. These materials exist mainly in the form of a plasma, a homogeneous mixture of ionized elements and their dissociated electrons. Solar density varies from around 10^5 kg/m^3 at the core (five times that of uranium) to 10^{-4} kg/m^3 at what we define as its surface (the *photosphere*). Solar radius ($\mathbf{R_s}$) at the photosphere is around 696,000 km. For comparison, the density of water is 1,000 kg/m^3, the mass of the earth ($\mathbf{M_e}$) is 6×10^{24} kg, and the radius of the earth ($\mathbf{R_e}$) is 6,378 km. The sun has a rotation rate which differs with solar latitudes—approximately 25 days per rotation at the equator and up

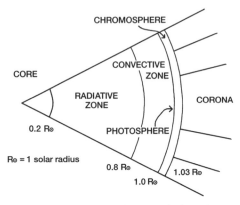

Figure 4-1. Solar regions. The sun's characteristics change with distance from its center.

to 30 days per rotation in the vicinity of the solar poles. The sun is composed of six major regions as shown in Figure 4-1.

Core. The core is the region of the sun in which the conversion of mass into energy occurs. Temperature in the core is about 15×10^6 °K and the density is about 110,000 kg/m^3, conditions sufficient to cause the fusion of hydrogen nuclei into helium nuclei with an associated release of energy. Approximately 4 billion kilograms of mass is converted into energy every second, producing about 4×10^{23} kW of power. There is enough fuel in the sun for this process to continue for at least a few billion more years, so there's no need to panic just yet.

Radiative Zone. In the region of the sun known as the radiative zone, the energy produced in the core is transported slowly upwards by means of absorption and reradiation of gamma rays, X-rays, and ultraviolet (UV) photons from layer to layer.

Convective Zone. Through the convective zone, the major transportation of energy is the *convection* (carrying) of energy by bubbles of hot gas that boil their way up to the surface. Through a telescope, these bubbles can be seen as a textured pattern on the sun called *granulation*. It is estimated that the transport of the energy created in the core takes about 10 million years to make its way through the radiation and convection zones to the photosphere, where it is released to the solar system.

Photosphere. The photosphere is a relatively thin layer (only about 300 km thick) of hot gas which absorbs the energy produced beneath it and then reradiates this energy into space. All electromagnetic radiation, including the visible wavelengths we see from earth, emanates from this region of the sun. Since this is what we perceive as its shape visually, this layer is defined as the "surface" of the sun even though the solar body extends far beyond. The temperature of the photosphere is roughly 6,000 °K which, as we will see, determines the characteristics of part of the solar radiation.

Chromosphere. The layer of gas above the photosphere is known as the chromosphere and corresponds roughly to the "atmosphere" of the sun. A temperature profile exists through this layer which initially decreases to about 4,000 °K then increases rapidly to about 10^6 °K (although temperature loses its familiar meaning with such large distances between particles). The density within the chromosphere is sufficiently low that the electromagnetic radiation from the photosphere is not appreciably affected in its outward passage through this region.

Corona. Finally, the corona is a layer of gases that extends millions of miles away from the sun. Temperatures remain high, and the density of the gases is on the order of 10^{-11} kg/m^3. The corona can be seen as a streaming pattern around the edges of the moon during a full solar eclipse.

Solar Radiation

The environment in the vicinity of the earth is affected by two major products of the sun's nuclear furnace. The most familiar is the electromagnetic radiation of which visible light is a part. The second is the outpouring of high energy solar particles.

Electromagnetic Radiation. Electromagnetic radiation is energy which propagates at the speed of light in a wave-like manner. Figure 4-2 depicts a typical sine-wave pattern used to describe an electromagnetic wave and illustrates some basic characteristics with which we will be dealing.

Wavelength (λ) represents the distance (meters, microns) between successive peaks of a particular wave pattern. Since electromagnetic energy travels at the speed of light, the number of peaks that pass a fixed point in a second of time is known as the *frequency* (f) of the radiation. The rela-

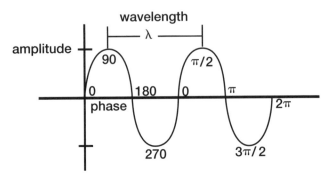

Figure 4-2. Electromagnetic wave. A simple sine-wave depicts some of the characteristics of an electromagnetic wave.

tionship between frequency and wavelength is given in equation 4-1 in which **c** represents the speed of light (3×10^8 m/sec in a vacuum).

$$f = \frac{c}{\lambda} \text{ cycles per sec ond (Hz)} \tag{4-1}$$

Two other characteristics of electromagnetic radiation, which we will look at more closely later, are the *amplitude* and the *phase*. Amplitude is represented by the height of the peaks of the sine-wave, and phase describes a particular location along the wave.

Figure 4-3 shows many of the regions of the electromagnetic spectrum with which we will be dealing. Notice how small the range of visible wavelengths is. Remember that as the frequency increases, the wavelength decreases, as explained by equation 4-1.

Blackbody Radiation. A *blackbody* is a theoretical body which has the properties of being a perfect radiator (and absorber) of energy. As we will see in the next few sections, the characteristics of the energy radiated by a blackbody is a function of the body's temperature. Though it is a theoretical concept, many bodies resemble blackbodies, at least closely or over a particular range of wavelengths, and can be approximated by blackbody relationships.

Stefan-Boltzmann Law. In 1879 an Austrian physicist, Josef Stefan, with help from his contemporary Ludwig Boltzmann, demonstrated that the total energy radiated by a blackbody increased as a function of the fourth

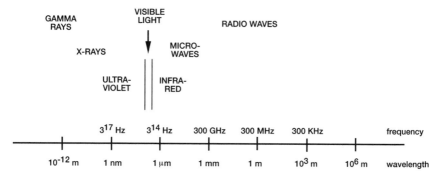

Figure 4-3. Electromagnetic spectrum. The major categories of electromagnetic radiations are indicated.

power of the body's absolute temperature. In terms of power output per unit area, the Stefan-Boltzmann relationship is:

$$E_{(T)} = \sigma T^4 \tag{4-2}$$

where the temperature **T** is in °K and σ represents the Stefan-Boltzmann constant ($\sigma = 5.76 \times 10^{-8}$ W/m² °K⁴). The temperature of the sun's photosphere is approximately 6,000 °K which, when substituted into the above relationship, results in a value $E_s = 74.7 \times 10^6$ W/m² indicating that each square meter of the sun's surface radiates more than 70 megawatts!

Solar Constant. We can estimate the total power output of the sun by multiplying the number just obtained from the Stefan-Boltzmann relationship by the surface area of the sun. Using the radius of the sun at the photosphere ($R_s = 696,000$ km), this total energy (power) can be found from:

$$P_s = E_s \times 4\pi R_s^2 \; (W) \tag{4-3}$$

Assuming negligible energy losses in space, the same amount of total energy must pass through any two spheres drawn around the sun. The earth circles the sun at an average distance of 149.5×10^6 km (represented by r_e, this has been adopted as a unit of distance measurement known as one *astronomical unit* or 1 A.U.). Using equation 4-3 to equate the energy emitted by the solar surface to that passing through a sphere of 1 A.U. radius, we get:

$$E_s \times 4\pi R_s^2 = E_e \times 4\pi r_e^2 \qquad (4\text{-}4)$$

where E_e represents the solar-generated energy per unit area in the vicinity of the earth. Of course, due to the increased surface area of the larger sphere, the energy in the vicinity of the earth is less than that at the surface of the sun. Using the value obtained for a 6,000 °K sun, we get a value of E_e = 1,631 W/m². This is the amount of energy that would impinge on a square meter area at a distance of 1 A.U. due to the energy output of the sun and is an important value when considering energy production via solar cells or for determining thermal inputs for a spacecraft.

This value (E_e) is known as the *solar constant*. Even though recent studies have shown that solar power output may fluctuate slightly (which may have affected the climate on earth severely at times in the past, perhaps contributing to the ice ages), for our purposes we shall consider it a constant value.

Wien's Displacement Law. In 1895, German physicist Wilhelm Wien, discovered that the wavelength corresponding to the maximum energy output for a blackbody at a particular temperature could be found from a simple relationship:

$$\lambda_{max} (\mu m) = \frac{2898.3}{T(°K)} \qquad (4\text{-}5)$$

For a temperature of 6,000 °K, equation 4-5 yields a maximum-energy radiated wavelength of 0.483 μm which corresponds to the yellow-green light frequencies close to the middle of the visible spectrum. The fact that human vision has adapted to take advantage of the maximum portion of the solar energy output seems like a good argument in support of the theory of evolution.

Planck's Law. In 1899, another German physicist, Max Planck, derived a relationship that combined the findings of his predecessors and described the distribution of blackbody radiation as a function of temperature and wavelength:

$$E_{(\lambda, T)} = \frac{2hc^2}{\lambda^5 [e^{(hc/KT\lambda)} - 1]}$$

where **K** = Boltzman's constant, **h** = Planck's constant, and the other terms are as described previously.

If this relationship is plotted for a temperature of 6,000 °K, the result would be as shown in Figure 4-4. The figure also shows the relationship plotted for a body at a temperature of 300 °K, which is the average tem-

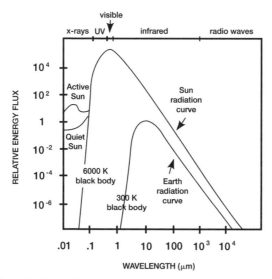

Figure 4-4. Planck's Law. The wavelength distribution of energy emitted by a body depends on its temperature.

perature of the surface of the earth. Both the earth and the sun behave as blackbody emitters over much of the electromagnetic spectrum (as we will see later). The sun emits energy at essentially all wavelengths of the spectrum with the maximum energy output in the visible frequencies as we determined earlier.

The energy levels for the earth are significantly lower (Figure 4-4 is plotted using logarithmic scales) and the peak of the curve is shifted to a higher wavelength (lower frequency). Due to its lower temperature, the earth radiates its maximum energies in the infrared wavelengths, as may be determined from equation 4-5.

One way to picture this temperature/wavelength relationship is to imagine a metal rod heated in a furnace. As its temperature is increased, the rod starts to glow, getting red then white hot, illustrating its radiation of different wavelengths with changing temperature.

Example Problem:

The sun's actual average "surface" temperature is about 5,750 °K. Based on this temperature, determine the blackbody power output of the surface of the sun, the expected solar constant in the vicinity of the earth, and the wavelength and associated frequency of maximum solar energy output.

Solutions:

From equation 4-2: $E_s = 62.96 \times 10^6$ W/m^2

Using equation 4-4: $E_e = 1,364.7$ W/m^2

From equation 4-5: $\lambda_{max} = 0.50$ μm

Finally equation 4-1: $f_{max} = 6 \times 10^{14}$ Hz

Solar Magnetic Field. The dynamic nature of the sun, causing motions of the solar plasma, generates strong magnetic fields. Magnetic field strength is commonly measured in units of Gauss or gamma (1 G = 10^5 γ) The strength and direction of the magnetic fields change over time and with changing processes on the surface, which will be described shortly. However, on average, the sun's surface magnetic field strength is about 15 G, approximately 30 times the surface magnetic field strength of the earth (0.5 G at mid-latitudes).

Particle Radiation. In addition to the electromagnetic energy released, there is a significant amount of particulate matter emitted into space by the sun of which there are basically two major sources: solar wind and solar flares.

Solar Wind. The solar wind, actually the extension of the corona, is a steady stream of plasma particles given sufficient energy to escape the sun's gravitational attraction and propagate into space in all directions. Through a process which implies that the product of the density and velocity of these particles remains constant with increasing distance from the sun, the velocity of this stream increases as it leaves the sun (to a point, after which it remains relatively constant).*

As mentioned earlier, the sun rotates while generating a substantial magnetic field. A characteristic of a plasma passing through a magnetic field is that the field lines become trapped and move with the plasma. As the solar wind propagates away from the sun, the solar magnetic field is dragged along but, due to its magnetic properties and the rotation of the sun, also affects the propagation of the plasma. The result is that the solar wind cannot be described as propagating in a straight path away from the sun. Actual propagation is three-dimensional and complex, making it difficult to predict the actual environment that may appear in the vicinity of the earth.

On average, the velocity of the solar wind in the vicinity of the earth is about 500 km/sec (supersonic in the definition of the term) and the density is about five particles/cm^3. Solar magnetic field strength is about 5 γ, approximately 1/10,000th the surface magnetic field strength of the earth.

As a final note, it is suspected that a boundary exists around 100 A.U. away from the sun where the streaming solar wind encounters the interstellar space environment defining a region surrounding the sun known as the *heliosphere.*† Some futuristic propulsion schemes propose to use the solar wind, like a sailboat uses atmospheric winds, to travel between the planets.

Sunspots/Solar Flares. The second major source of particle radiation is due to disturbances on the solar surface. The processes which produce sunspots and solar flares are intimately linked and must be described together.

Sunspots. Galileo, using the crude telescopes of his own design, was the first to notice "blemishes" or spots on the surface of the sun. Even though he had no idea of the processes that may have accounted for these spots, he did keep a close enough watch on them to determine that the sun

---

*The *Ulysses* spacecraft, a joint NASA/ESA mission launched from the Space Shuttle in October 1990, used a gravity assist from Jupiter to change its orbital plane out of the ecliptic (the sun's equatorial plane within which the planets more or less lie) to sample the interplanetary space environment over the sun's poles. The results show a difference in the velocity/density relationship from the ecliptic, which has yet to be fully explained.

---

†*Pioneer 10* (launched in December, 1973) and *Voyager 1* and *Voyager 2* (launched in September 1977 and August 1977, respectively) interplanetary spacecraft are on trajectories that will carry them beyond the solar system. In 1996, *Voyager 2* will be around 45 A.U. from the sun, and *Pioneer 10* and *Voyager 1* will both be approximately 60 A.U., all traveling in different directions. These still-operating spacecraft continue to be monitored for information to help determine the extent and characteristics of the heliosphere.

rotates on its axis, which we now know is a rotation rate which differs with solar latitudes. As it turns out, this differing rotation rate represents part of the process which produces sunspots and solar flares.

As mentioned earlier, magnetic fields passing through a plasma become trapped and move with the plasma. The differing rotation rate of the sun causes these trapped magnetic field lines to "wind up" and increase their field strength to the point where they produce a magnetic pressure on the plasma. In these areas, the gas pressure is lower, which lowers the gas temperature. As we have seen, the sun "looks" yellow due to the temperature at which its surface is radiating. Sunspots are darker-looking patches of lower temperature due to these areas of increased magnetic field strength at the surface. Magnetic field strengths of 1,500 G in these locations are not uncommon.

Solar Flares. The actual cause of solar flares is still not completely understood. One explanation states that the magnetic field lines continue to wind up to a strength where they pop out from the surface of the sun, carrying solar plasma into space. The escaping material may be accelerated by the field above the surface to velocities approaching relativistic speeds up to 10^6 m/sec. Intense electromagnetic energy in all wavelengths, from X-rays through radio, is also released by solar flares. As we shall see shortly, arrival of these high-energy particles and intense radiation to the vicinity of the earth creates a situation which can affect the operation of ground systems as well as spacecraft.

Solar Cycle. Records have been kept on the number of sunspots visible on the sun's surface since the mid-1700s, and review of these records indicates that the number of sunspots changes over time with some regularity. At times, no sunspots at all may be visible, while at other times over 100 sunspots may exist simultaneously. If one describes the period of little or no sunspot activity as *solar minimum* and the period of highest activity as *solar maximum,* the average time from one minimum or maximum period to the next can be defined as the *solar cycle.* Though the solar cycle varies between approximately seven to thirteen years, the average cycle has been found to be about eleven years. Also, on average, it seems to take less time to rise to solar maximum, around four years, than to settle back to solar minimum, around seven years. The next solar maximum is expected to occur around the year 2000.

Sunspots seem to occur only around mid-solar latitudes, between around 60° north and south of the equator, forming closer to the equator as the cycle approaches minimum. A reversal of the magnetic polarity of the sun, from positive to negative lines of magnetic flux emanating from the geographical poles, occurs over each solar cycle. This variability in intensity, location, and magnetic polarity over time manifests itself in a changing and difficult-to-predict environment in the vicinity of the earth.

THE EARTH

The environment on and around the earth is affected in many ways by the electromagnetic and particulate radiations from the sun. Spacecraft must operate in and through the resulting environment, which includes the earth itself and its atmosphere as well as the space around it. Consequently, the following sections describe some of the properties of the earth as well as the interaction of the earth with the solar radiations.

Electromagnetic Interaction

We previously determined the amount of electromagnetic energy that reaches the vicinity of the earth when we calculated the solar constant. This energy retains the blackbody spectral characteristics it had when it left the sun, the only difference being that the energy levels over the range of wavelengths has decreased. As this energy encounters the earth, the interactions affect both the earth and the energy itself. In some cases, these interactions are useful to the space systems user. For example, remote sensing techniques measure the absorption or transmission of solar radiation through the atmosphere to infer some of the characteristics of the atmosphere from space, and most spacecraft use solar panels to convert some of this radiation to electricity to power the spacecraft systems. In other cases, solar radiations produce undesired effects. For instance, spacecraft may absorb excess infrared frequencies, necessitating the use of thermal control devices, and the ionosphere, described next, may introduce errors in transmitted signals. Spacecraft systems, including power generation and thermal control, and spacecraft applications, such as remote sensing and communications, are described more thoroughly later in the text.

The Ionosphere. The short wavelength (X-ray, UV) radiations from the sun have enough energy to "knock" electrons out of some of the constituents of the atmosphere at high altitudes. This process is called *photoionization* and results in an electrically charged (ionized) region known as the *ionosphere*. This region begins somewhere between 50 and 70 km altitude and extends far into space. The ionosphere is characterized by the changes in the free electron concentration levels with altitude as shown in Figure 4-5.

Figure 4-5 also shows the variability of the ionosphere. Solid lines indicate levels during times of high solar activity and dashed lines are associated with low solar activity levels. Notice also the considerable difference between day and night characteristics. This is due to the cessation of the photoionization process when out of direct exposure to the sun. During these times the electron concentration is reduced due to the recombination of ions and electrons. This process proceeds slowly enough to allow a concentration of free ions and electrons to persist throughout the night.

Some easily identifiable regions within the ionosphere can be seen in Figure 4-5 associated with changes in the electron concentration with altitude. These regions (layers) are identified with letters (D, E, F_1, F_2, or just F). This layering of the ionosphere is associated with the changes in atmospheric constituents with altitude which have differing ionization properties. We will return to our discussion of the ionosphere later when we discuss communications, remote sensing, and navigation satellites.

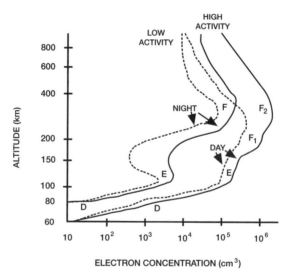

Figure 4-5. The ionosphere. Electron concentration in the ionosphere changes with altitude, time of day, and solar activity.

The Atmosphere. A chapter on space environment may seem like a strange place to include a discussion of the earth's atmosphere, but the atmosphere is a factor in the operation of spacecraft in many ways. The following sections describe the atmosphere, along with its interaction with the sun's electromagnetic radiations.

Composition of the Atmosphere. It is believed that the atmosphere was created as a result of the exhausting of materials from the interior of the earth by volcanic activity. Nitrogen, being relatively inert, remains as the most common atmospheric element. Diatomic oxygen was mainly produced via the photosynthesis of water and carbon dioxide by plant life over millions of years. Argon and other trace elements make up the remainder of the earth's atmosphere as shown in Table 4-1.

Table 4-1
Composition of the Earth's Atmosphere (below 100 km)

Constituent	Content (fraction of total molecules)
Nitrogen (N_2)	0.7808 (75.51% by mass)
Oxygen (O_2)	0.2095 (23.14% by mass)
Argon (Ar)	0.0093 (1.28% by mass)
Water vapor (H_2O)	0–0.04 (variable)
Carbon dioxide (CO_2)	325 parts per million
Neon (Ne)	18 parts per million
Helium (He)	5 parts per million
Krypton (Kr)	1 part per million
Hydrogen (H)	0.5 parts per million
Ozone (O_3)	0–12 parts per million

Higher in the atmosphere, monatomic oxygen (single oxygen atoms) are created through *photodissociation* (splitting) of molecular oxygen by the sun's radiation. Monatomic oxygen is highly reactive, and some of these atoms combine with diatomic oxygen to produce the ozone (O_3) found in the atmosphere. Ozone is particularly important to life on earth because it absorbs much of the harmful ultraviolet radiation emitted by the sun.

Temperature Variation. The variation of temperature with altitude up to about 700 km through the atmosphere is shown in Figure 4-6. These values are for a "standard" atmosphere which represents typical conditions at

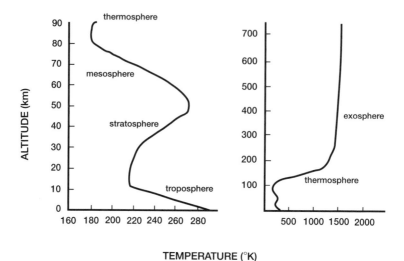

Figure 4-6. The "standard" atmosphere. The temperature of the atmosphere varies with altitude.

mid-latitudes. This temperature variation delineates several distinct regions also identified in the figure.

Troposphere. The troposphere is an area of unstable air where most of the weather we experience is formed. The region is unstable due to the turbulent convection of air heated at the surface by the infrared energy that the earth emits due to its temperature, which is a result of the earth's absorption of solar radiation. As can be seen, the temperature decreases linearly through this region until we reach the tropopause—the transition region between the troposhere and the stratosphere—where the temperature remains relatively constant (isothermal).

Stratosphere. Unlike the troposphere, the stratosphere is relatively nonturbulent. Temperature increases with altitude through this region due to the continued absorption of the earth's infrared radiation by H_2O and CO_2 molecules and absorption of solar ultraviolet radiation by the ozone that exists at the top of this region.

Mesosphere. Above the isothermal layer at the top of the stratosphere (the stratopause), the mesosphere is a region where the temperature once again decreases with altitude. This is due to the reradiation of energy absorbed from the layers below which escapes into space.

Thermosphere. After the mesopause (another isothermal layer), the thermosphere represents a region in which the temperature increases rapidly with altitude. This is due, once again, to the direct molecular absorption of solar radiation by the atmospheric constituents at these altitudes.

Exosphere. In the exosphere, temperature remains relatively constant with altitude due to the low densities and the large mean free path (distance) that exists between molecules. In this region, energetic molecules may actually escape from the earth's gravitational pull and be lost into space.

Density Variation. Decrease in the density of the atmosphere with increasing altitude is a result of the balance between the gravitational force on molecules of different masses and the thermal energy of these molecules. Though relationships have been developed that relate the change in density with altitude based on the standard temperature profile, they do not work well at the higher altitudes at which spacecraft operate. A good rule of thumb is that up to 100 miles altitude, the density decreases by a factor of 10 every 10 miles. Above this, the decrease becomes exponential as the lighter elements such as hydrogen and helium become more predominant. The atmosphere extends for thousands of miles above the earth's surface, but in ever-decreasing densities.

The density of the atmosphere at any particular altitude also varies with time due to the variability of solar emissions over the solar cycle and during solar flares. Increased emissions may heat up the atmosphere causing an increase in density at satellite altitudes which increases drag and may affect orbital lifetimes.

Electromagnetic Propagation Through the Atmosphere. As the previous discussions have indicated, the atmosphere absorbs much of the electromagnetic energy radiated from both the sun above and the earth below. It also reradiates much of this energy in both directions as well, but the average rates of absorption and emission are about equal, which results in a relatively constant average temperature around the globe.

The mechanisms behind absorption and emission are rooted in chemistry, physics, and quantum theory which show that these properties happen at relatively distinct frequencies for different atoms and molecules.

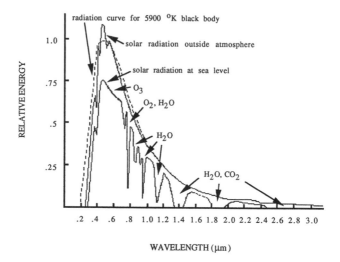

Figure 4-7. Electromagnetic propagation. Much of the solar radiation incident on the earth is absorbed by the atmosphere.

Figure 4-7 shows the effects of this compatibility of certain elements with certain frequencies (wavelengths) over the range of radiations incident on the earth from the sun.

The topmost curves in Figure 4-7 represent the spectral energies that arrive from the sun at the top of the earth's atmosphere. The dashed line indicates the theoretical energies given by the relationships discussed earlier in this chapter for an idealized blackbody. As you can see, the correlation of this theoretical curve with the measured irradiation outside the atmosphere is quite good. The lower solid curve indicates the amount of energy, by wavelength, that reaches the earth's surface. Although there is attenuation in most of the incident radiations, you can see that there are some wavelengths which are drastically (and some completely) attenuated in their passage through the atmosphere. The associated elements which absorb these energies are also indicated on the figure.

The Greenhouse Effect. Figure 4-7 shows that much of the energy of the sun's radiation does reach the earth's surface. This incident radiation warms the surface, which then reradiates this energy—due to this temperature and Wien's law—in the infrared frequencies. However, the CO_2 and H_2O molecules in the atmosphere are highly absorbent in the infrared and, as a consequence, the atmosphere is heated, as was indicated in the earlier descriptions of the atmospheric regions. As it stands, there is a thermal balance between the amount of energy absorbed and the amount reradiated into space, but the above discussion shows the concern over increasing the carbon dioxide levels in the atmosphere as can happen through the large-scale burning of fossil fuels and forests.*

Particle Interactions

To discuss the characteristics of the solar particle radiations in the vicinity of the earth, we must first discuss the earth's own magnetic field and the interaction of charged particles with this field.

Geomagnetic Field. As any good scout knows, the earth has a magnetic field on which the pointer of a compass can align and direction be determined. One explanation for its existence is given by the Dynamo Theory which states that the field is generated by a complex and nonuniform rotation of the earth's molten metallic core. As shown in Figure 4-8, the *geo-*

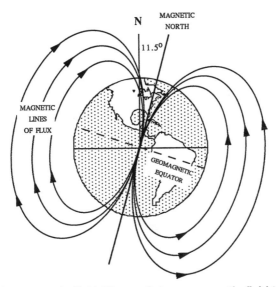

Figure 4-8. Geomagnetic field. The earth has a magnetic field that can be modeled as a huge magnet located near the earth's center.

*It is believed that high atmospheric CO_2 concentrations are a major contributor to the 700°F conditions on the surface of Venus—temperatures hot enough to melt lead!

magnetic field can be modeled by a (quite large) dipole magnet placed within the earth and tilted about 11.5° from the earth's axis of rotation. This theoretical magnet would not reside exactly at the geometric center of the planet but would be displaced about 436 km off-center in the direction of the Pacific Ocean.

Interaction of the Solar Wind and the Geomagnetic Field. When the solar wind encounters the Earth's magnetic field, an electromagnetic interaction occurs that deforms the geomagnetic field as depicted in Figure 4-9. As the supersonic charged particles of the solar wind encounter the magnetic field of the earth, a *shock front* is formed across which the solar particles are deflected and slowed, similar to that which occurs in aerodynamics. The distance to the earth of the shock front is about 14 earth radii (14 R_e) at its closest in the direction of the incoming solar wind, increasing greatly as the angle away from this direction increases.

Inward from the shock front, the particles exchange kinetic energy for magnetic energy, which is what causes the geomagnetic field to deform. The *magnetopause* represents the boundary where the magnetic forces overcome the dynamic forces of the solar wind. This is a relatively stable boundary requiring an increase in solar kinetic energies, such as occurs with solar flares, to change significantly. The magnetopause is closest at

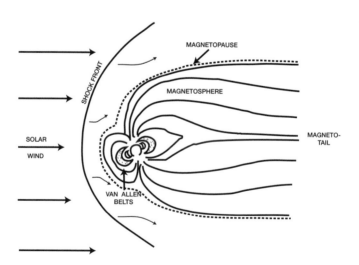

Figure 4-9. Solar wind. The earth's magnetic field is distorted with its interaction with the solar wind.

about 10 R_e, but the tail formed "behind" the earth extends far beyond the orbital distance of the moon ($r_{moon} = 60$ R_e) Within the magnetopause rests the *magnetosphere* which defines the cavity containing the (deformed) geomagnetic field.

Geomagnetic Storms. The overall kinetic energy of the solar wind is increased significantly during times of increased solar activity (increases in sunspots and solar flares). When this additional ejected plasma (at speeds up to 10^6 km/sec) encounters the earth, it causes a fluctuation of the geomagnetic field as illustrated in Figure 4-10.

From several minutes to several hours after a solar disturbance, the additional high-energy particles arrive, causing a compression of the earth's magnetic field measured as a sharp increase in the field strength at the earth's surface. This sudden commencement phase of the geomagnetic storm is followed by a two- to eight-hour period during which the magnitude of the field strength remains high. After this initial stage, when the solar wind returns to its prestorm intensity, the geomagnetic field drops to an intensity level significantly lower than normal due to currents set up by increased motions of charged particles trapped within the field, as described in the next section. This main phase lasts about 12 to 24 hours after which a recovery phase begins, lasting typically a few days, during which the field gradually dissipates the excess charges and returns to its normal level of intensity.

The drastic alteration of the earth's magnetic field associated with geomagnetic storms can affect the operation of many space and terrestrial

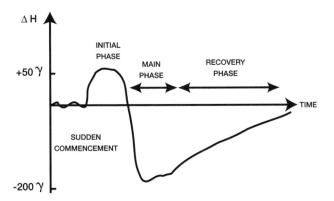

Figure 4-10. Geomagnetic "storm." The intensity of the earth's magnetic field may be greatly affected by major solar disturbances.

systems. The ionosphere may also be altered, which can cause interruption of normal communication channels. The increased intensity of charged particles can affect unprotected systems in space and on the ground, causing overloads, processing or other electronic errors, and increased radiation hazards to humans in space. An associated increase in the motions of trapped charged particles, described next, can also cause an increase in atmospheric density which may affect satellite lifetimes.

Radiation Belts

The presence of areas of radiation around the earth was discovered by Dr. James Van Allen using instruments aboard the early U.S. V-2 test flights and, later, Explorer series satellites. It was found that some of the charged particles present in space were able to enter the magnetosphere and become trapped within the geomagnetic field. The gain/loss mechanics of this phenomenon is still not completely understood, but the levels, dispersion, and behavior of the trapped particles have become well described.

Early representations of "belts" of radiation surrounding the earth are better described by distributions of particles within the magnetosphere which follow closely the geomagnetic lines of flux. Areas of different concentrations exist due to the types of particles (protons and electrons) and associated energy levels. More energetic particles are generally trapped closer to the earth, and the major concentrations of particles occur around the equator where the minimum value of magnetic field flux occurs. Figure 4-11 shows this distribution of trapped particles around the earth in terms of particle types and energy levels (1E2 indicates 1×10^2 MeV or mega-electron volts of energy). In some regions, the level of radiation due to these trapped particles is sufficient to be disruptive to spacecraft operations and hazardous to humans.

Although caught in the earth's magnetic field, these particles are far from stationary. Motion of these energetic charged particles is a combination of spiraling around the magnetic field lines while bouncing from pole to pole along the lines. A slower drift around the equator also occurs. When these particles interact with the ionosphere as they spiral in towards the earth in the vicinity of the poles, they produce the auroral phenomena known as the "northern lights" in the northern hemisphere. A more serious effect of these motions is the contribution to atmospheric heating by particle precipitation in the auroral zones and by the electric currents flowing at high and low latitudes.

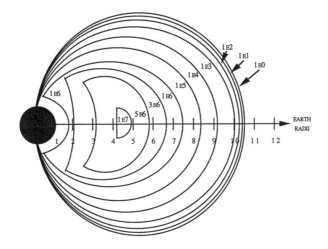

1 E2 INDICATES 1 x 10² MEV ENERGETIC PARTICLES

Figure 4-11. Radiation belts. Energetic particles are trapped within the earth's magnetic field.

South Atlantic Anomaly. The off-set geometry of the geomagnetic field, shown in Figure 4-8, produces an asymmetry in the radiation belts which results in an area where the belts are closer to the earth's surface. This area occurs just off the coast of Brazil in the Atlantic Ocean and is known as the South Atlantic Anomaly. Spacecraft passing over this area experience a significantly increased radiation dosage sufficient to cause concern when operating in this region.

SPACECRAFT EFFECTS

The space systems engineer must not only be familiar with the environment in which a system is to operate, but must also determine how that environment is going to affect the spacecraft physically and operationally. The following sections introduce some of the major concerns. In most cases, a spacecraft must have systems or operational methods to deal with these effects.

Thermal Control

Spacecraft in orbit receive thermal inputs from many sources including direct exposure to solar radiation, infrared emissions from the earth, and heat generated by onboard systems. Many of these sources may cycle on and off, such as the solar inputs when orbiting from dark to light or from spacecraft systems as they turn on and off. Spacecraft also continuously emit heat, especially toward the "cold" (absorptive) blackness of space. For some applications, certain spacecraft parts may always point toward the sun while others may always point toward space. Within certain defined limits, the craft must be able to balance the temperatures of all its components. Several spacecraft have been lost or impaired due to the freezing of propellant lines or the heating of an electronic component to the point of failure.

Spacecraft Charging

As we have seen, the space environment is full of charged materials, trapped solar particles, and ionized atmospheric constituents as the major examples. Also, as they travel through space, orbiting spacecraft cut through magnetic field lines of flux which tends to make charges want to move. Operation within such an electrically rich and magnetically dynamic environment can result in a buildup of charges on the surfaces of the spacecraft to the point at which a current flow or discharge from one component to another may result in damage to the system.

Radiation

Some of the higher-energy particles present in space can penetrate the skin and components of a spacecraft. These may deposit an electrical charge inside electrical components or may even change the physical structure of materials. Sudden depositions of electrical charge in computer logic circuits or memories can corrupt data or even disrupt operation of the satellite if a "phantom command" is sensed from such an event. Buildup of radiation damage to semi-conductor materials can cause computer chips to degrade to the point where commands may no longer be able to be processed. Satellite operators have to be especially vigilant for signs of these effects during periods of high solar activity.

Vacuum/Corrosion

As we have seen, the near-earth space environment is not a total vacuum. A spacecraft in orbit is constantly being bombarded by atmospheric particles moving at relatively fast velocities. Some of these particles are the monatomic oxygen atoms created by photodissociation. Being a highly reactive element, when these atoms strike a spacecraft they tend to combine with the spacecraft materials, creating a potentially damaging corrosion over long periods of time in space.

The major problems associated with operating in the *near*-vacuum environment of orbits around the earth concern the materials used. Some materials may change their physical properties over time through *outgassing*—the release of trapped gases within the materials—or from the radiation and/or corrosion damage discussed earlier. Materials in contact in a vacuum can actually transfer atoms across their surfaces and eventually "cold weld" to each other. This may cause severe problems in spacecraft with moving parts. Special substances must be used to separate and lubricate moving parts in space, because lubricants commonly used on earth would simply boil away in a vacuum.

Micrometeorites

The earth is constantly being bombarded by small particles from space. Meteor showers occur when these particles encounter the atmosphere and burn up with the friction of reentry. Though extremely small, these particles may have enormous velocity differences with an orbiting spacecraft. Collision with these micrometeorites can pit the surface of a spacecraft and degrade the performance of sensors or systems (such as solar power cells and thermal coatings) and perhaps cause more serious damage.

Man-made Debris

In the short time that man has been able to place objects in space, the debris created by launch vehicles and spacecraft operations has already created a situation of some concern. The U. S. Space Command has identified more than 7,000 artificial space objects of 10 cm diameter or larger orbiting the earth. Almost 6,000 of these are in low-earth orbit, and about half of the count are spent rocket stages and active and inactive satellites. Computer simulations predict approximately 17,500 objects between 1 and 10 cm in diameter along with millions of smaller particles of space debris. The total mass of man-made debris is about 15,000 times the naturally occurring mass concentration around the earth. The relative velocity between objects that may collide in low-earth orbits is around 10 km/sec

(22,000 mph). Such hypervelocity collision with particles as small as 1 cm in diameter can cause significant impact damage, and shielding for particles much larger than this becomes impractical. This subject has become one of great concern in the design of the international space station. It is estimated that at least one impact with an object 1 cm in diameter or larger will occur every 2.5 years for a space station sized vehicle, and about 50,000 impacts with particles 0.01 cm or less would occur each year.

Drag

As mentioned earlier, the earth's atmosphere extends thousands of kilometers above the surface, albeit in thinner and thinner proportions. Besides the potentially damaging interactions with atmospheric elements, the drag on spacecraft in this region is also not negligible and may affect spacecraft lifetimes and attitude control. This was graphically exemplified by the reentry of the huge *Skylab* space station after only six years in orbit. Solar effects on the atmosphere (such as from solar flares and changes in the amount of UV radiations) can cause changes in the atmospheric density, affecting the drag on orbiting spacecraft dramatically and unpredictably.

Sensors

Sensors used in spacecraft are usually very sensitive to particular radiations. Direct exposure to solar radiation may result in unwanted signal reception and possible sensor damage. If the purpose of a system is to pick up the weak radiations of a faraway planet or star, and the sensor happens to inadvertently "look" in the direction of the sun (or a reflection of some of its energy), the desired signals may be obscured in the sun's radiations and the sensor itself may be damaged. Additionally, the sensors must be protected from the general effects of the space environment described above or else performance may be adversely affected.

Spacecraft Performance

The operability and performance of a spacecraft may also be affected by the space environment *through* which it must operate. Remote sensors must peer through an atmosphere of changing properties (heat, cold, rain) and conditions (night, day), and the transmission of information to ground stations is affected by the atmosphere and ionosphere which also fluctuate constantly.

Man in Space

All of the above situations must be considered when man is placed into the space environment. In most cases, a suitable environment can be provided within a spacecraft or with the use of a space suit. However, humans are particularly susceptible to high-energy particle radiations in the form of cosmic rays (from distant stars and galaxies), the particles trapped in the geomagnetic field, and those produced by solar flares. These particles have the ability to pass through shielding and can produce cell damage to astronauts, so exposure must be measured and limited. Fortunately, the areas associated with commonly used nonpolar low-earth orbits have low enough radiation levels to be relatively safe with simply a small amount of protection.

REFERENCES/ADDITIONAL READING

National Aeronautics and Space Administration, *Space Physics Strategy—Implementation Study,* 2nd ed. Washington: NASA, 1991.

Interagency Group (Space), National Security Council, *Report on Orbital Debris.* Washington: National Security Council, 1989.

T. Wilkerson, M. Lauriente, and G. Sharp (Eds.), *Space Shuttle Environment.* New York: Engineering Foundation, 1985.

C. Cochran, D. Gorman, and J. Dumoulin (Eds.), *Space Handbook.* Alabama: Air University Press, 1985.

Tascione, T., *Introduction to the Space Environment.* Colorado Springs: U.S. Air Force Academy, 1984.

EXERCISES

1. Draw a graph showing the estimated temperature (T) and density (ρ) variations of the sun from 0 to 2 solar radii (R_s) based on the information given in the chapter. Use logarithmic scales for T and ρ and a linear scale for R_s.

2. Starting with a value for the solar constant in the vicinity of the earth of 1,390 W/m^2:

a. Compute the amount of power that must be radiated from the surface (photosphere) of the sun (E_s W/m^2) to produce this value at the earth.

b. Compute the total power output of the sun (W).

c. Using the figures given in the reading and Einstein's energy/mass relationship, determine the amount of hydrogen being converted to helium to produce this power (kg/sec).

d. Compute how long it will take until the sun uses up half of the remaining hydrogen.

e. Using the Stefan-Boltzmann relationship, determine the corresponding surface temperature of the sun to produce the power generated.

f. Using Wien's displacement law, compute the maximum radiated wavelength associated with this temperature.

3. If the earth were a blackbody with an average temperature of 300 °K, determine the wavelength of maximum radiation emitted (μm), the power emitted by the surface of the earth (W/m^2), and how much of this power would impinge on a spacecraft orbiting at geostationary altitude (W/m^2). ($r_{geosta} = 42,164$ km.)

4. Some objects in space are huge emitters of ultraviolet radiation. If a UV frequency of 3×10^{16} Hz were detected from a suspected blackbody, determine the temperature and the amount of power emitted by the surface of this body.

PART 2

Spacecraft Applications

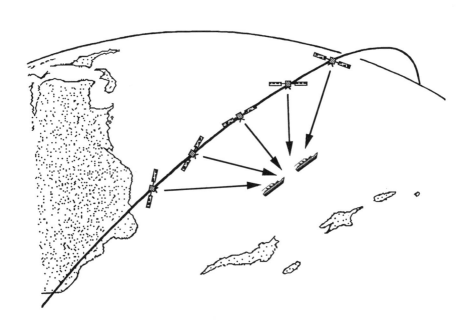

CHAPTER 5

Communications

One of the first practical uses for systems permanently stationed in space was proposed by science fiction writer Arthur C. Clarke (most commonly known for writing the story behind *2001: A Space Odyssey)* over ten years before the first crude satellites achieved orbit. Having a knowledge of orbital mechanics, Clarke realized that a satellite in a geostationary orbit could connect, in many ways, any two points that might fall within the large field of view. In the early days of telephone, it might not have seemed too unusual to imagine a switchboard in the sky allowing easy, instantaneous communications between such locations. Though devoid of operators, hundreds of such switchboards exist today and connect millions of places around the globe from the geostationary band above the earth's equator.

COMMUNICATIONS THEORY

Figure 5-1 reviews some of the terms that describe the characteristics of electromagnetic waves as discussed in Chapter 4.

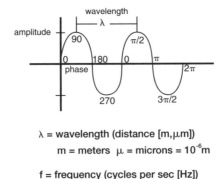

λ = wavelength (distance [m,μm])

m = meters μ = microns = 10^{-6}m

f = frequency (cycles per sec [Hz])

= c/λ c = 3 x 10^{8}m/s

Figure 5-1. Electromagnetic wave. The characteristics of an electromagnetic wave are used to carry information for communications.

This chapter concerns electromagnetic amplitude and phase as well as frequencies and wavelengths. *Amplitude* represents the amount of energy or power the wave contains, and *phase* indicates where along the wave form we are at any particular time (in degrees or radians as shown).

As we will see in this section, electromagnetic waves are used for communication purposes to carry information between locations. Before describing this process, however, some additional information on electromagnetic propagation must be presented. These properties and phenomena also apply to remote sensing and satellite navigation, discussed later.

Radio Wave Propagation

The electromagnetic spectrum was described in the last chapter in terms of different bands of frequencies with certain properties. The portions of the spectrum we will be concerned with here are the *radio* and *microwave* bands which are most commonly used for communication purposes. Figure 5-2 shows the frequencies associated with these two bands and also shows how these frequencies have been further subdivided into bands

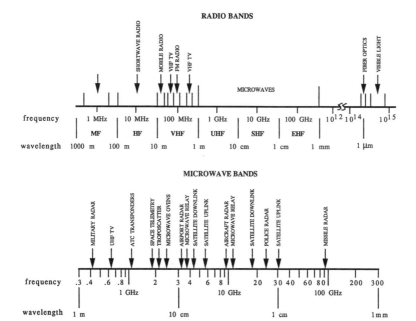

Figure 5-2. Communication bands. Certain frequencies have been designated for particular communications uses.

given names like VHF and UHF. Through international agreements, certain frequencies have been assigned to particular uses, such as AM broadcast and shortwave. These uses have come about mainly due to the propagation properties of the frequencies and their usefulness in carrying certain types of information. These points will be further explained in the following sections.

Figure 5-3 shows how electromagnetic waves of certain frequencies propagate differently. These propagation characteristics and the reasons behind each are explained as follows:

Direct Waves. Essentially any frequency can be transmitted between two stations with a direct, unimpeded path between. This is known as *line of sight* (LOS) communications and is used in many practical applications using frequencies in the radio, microwave, and even visible and other electromagnetic bands.

Ground Waves. At frequencies less than a few megahertz, electromagnetic energy can interact with the material in the earth and tend to follow the contour of the earth's surface. This property allows stations to communicate *over the horizon* (OTH) in longer ranges than direct path stations. As we will see, however, these lower frequencies are limited in how much information they can transmit. Frequencies associated with ground waves are commonly used for public *amplitude modulated* (AM) broadcasts. *Frequency modulated* (FM) radio stations broadcast on higher frequencies that require direct paths between the transmitter and receiver. This is the reason why AM stations can be picked up generally farther away than FM stations. (Note: Modulation techniques are presented later in this chapter.)

Sky Waves. Frequencies higher than AM frequencies lose the ability to follow the surface contour but, up to a certain point, can have an interaction with the ionosphere which allows longer range propagation.

Ionospheric Interaction. The ionosphere was described in Chapter 4 as a region of electrically-charged (ionized) atmospheric constituents. Electromagnetic waves, upon encountering these electrical charges, tend to change their direction of propagation through *refraction.* If enough charges are encountered, the wave can be turned back downward toward the surface, appearing to be "reflected" by the ionosphere at a certain altitude. Depending on the wave frequency and the electron concentration of the ionosphere at the time, some waves will propagate farther into the ionosphere before being turned, appearing to be reflected at a higher altitude.

Knowing the ionospheric makeup at any time, one can determine the frequencies for which refraction of a signal will occur. The critical frequency for this type of interaction is given by equation 5-1.

$$f_{crit} = 9 \times 10^3 \, (N_e)^{1/2} \, (Hz) \tag{5-1}$$

where N_e represents the electron density (electrons/cm^3) at a particular region of the ionosphere. Equation 5-1 reveals the *highest* frequency that will be reflected (with *normal* incidence) by a particular region of the ionosphere. Lower frequencies will also be reflected, but higher frequencies will not be refracted sufficiently to be turned completely back down toward the surface.

As an example, from Figure 4-5 in Chapter 4 we can see that the highest electron density for the lowest region of the ionosphere (the "D" region) is about 10^3 electrons/cm^3. From equation 5-1, the highest frequency that this layer will reflect is about 280 kHz. We can determine the highest frequency that will be reflected by the ionosphere as a whole by noticing that the highest electron concentration indicated in Figure 4-5, associated with the F$_2$ region, is about 2×10^6 electrons/cm^3. Equation 5-1 indicates that frequencies up to about 13 MHz will be reflected somewhere in the ionosphere, and frequencies higher than this may pass through these layers and into space. Note that these results were achieved for the average daytime electron concentrations as given in Figure 4-5. Actual concentrations may be slightly different and, of course, conditions change significantly at nighttime. Both these situations affect the critical frequencies and show why continual monitoring of the ionosphere is important.

As noted above, equation 5-1 only addresses a signal propagating normal to the ionospheric layers. A signal that encounters the ionosphere at some angle ϕ, as depicted in Figure 5-3, will encounter more charged particles as it passes through the regions and will have a better chance to be refracted. As a result, higher frequencies than those obtained for normal incidence will be turned depending on this angle, as can be determined from equation 5-2.

$$f_{crit}(\phi) = \frac{9 \times 10^3}{\cos \phi} \, (N_e)^{1/2} \, (Hz) \tag{5-2}$$

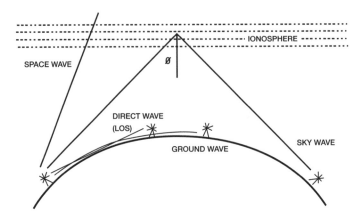

Figure 5-3. Electromagnetic wave propagation. Certain frequencies of electromagnetic energy propagate differently.

The ranges achieved depend on a number of factors including frequency, angle of incidence, and transmitted power (to be discussed shortly), as well as time of day (or night) and other factors which may affect the propagation of the signal. A powerful enough signal may, after refraction by the ionosphere, reflect off the earth's surface back toward the ionosphere to be refracted again. These multiple-hop transmissions can achieve tremendous ranges and allow essentially worldwide communications. Frequencies used for this type of communication make up the *shortwave* band which has been used for long-range communications since Marconi explored the possibility of such use at the beginning of this century.

Space Waves. Above the critical frequency (generally above 30 MHz), electromagnetic waves are not affected by the ionosphere enough to be deflected back down to the surface. Communications using these frequencies are all line-of-sight (LOS) communications because the transmitting station and receiver must have a direct path between the two.

Antenna Theory

Electromagnetic waves that propagate through space originate from, and are received by, some sort of antenna structure. Power is delivered to an antenna at the desired frequency, causing electron movement in the structure. The movement of electrons causes an associated electromagnetic wave to be formed which propagates from the antenna in a certain pattern depending on the physical characteristics of the antenna.

Antenna Patterns. An antenna pattern is a geometrical plot that represents the way in which an antenna propagates electromagnetic energy. In most cases, the pattern is the locus of points that represent the range at which the transmitted power is decreased decreased equivalently by 3 decibels (dB) of its original value (or some normalized equivalent value). The relationships between *decimal* values and *decibel* values are given by:

$$x \text{ dB} = 10_{\log_{10}}(x) \tag{5-3}$$

and

$$x = 10^{\frac{x \text{ dB}}{10}} \tag{5-4}$$

The second of these equations shows that −3 dB equals one-half in decimal value. This indicates that the locus of −3 dB points of the antenna pattern represent the *half-power* locus delineating where the signal has decreased to one half of its originally transmitted power level.

Isotropic Radiator. An isotropic radiator is a (theoretical) structure from which electromagnetic radiations would propagate in all directions at equal power strengths. The antenna pattern for such a radiator would look like a perfect sphere of half-power points emanating uniformly in all directions from a central point.

Anisotropic Radiators. In many cases, it is desired to concentrate transmitted radiations in a particular direction or pattern in order to reach certain areas and/or certain power levels in those directions. All practical antennas are designed, depending on their specific purpose, to transmit (and receive) signals in such a way. For instance, the radio antenna on your car must be able to receive AM and FM signals from nearby radio stations in any horizontal direction (since you may be making turns as you drive), but because no stations currently exist in space or underground, the antenna does not need to pick up signals from these directions. The familiar pole antenna on your car has an antenna pattern that looks like a doughnut (when looking from above or below), indicating that the antenna concentrates its attention horizontally with decreased receiving capabilities in other directions.

Parabolic Dish. In many applications, it is desired to concentrate a signal as much as possible in a single direction. For these applications, a parabolic dish antenna is most commonly used. Figure 5-4 depicts a parabolic dish with a representative antenna pattern, and also delineates some of the terms with which we will be dealing.

When transmitting, the signal is either emitted by, or reflected off, a structure at the *focal point* of the antenna paraboloid. The parabolic shape of the antenna reflects the signal mainly along the *boresight,* concentrating most of the signal power in this direction. When receiving, the incoming electromagnetic wave "reflects" off the dish and is concentrated onto the focal point of the parabola where (either by pickup or secondary reflection) the signal is delivered to the receiving components. *An antenna has the same propensity for receiving a signal from a certain direction as it has for transmitting in that direction,* as quantified in the following section.

Antenna Gain. The ability for an antenna to concentrate a signal in a certain direction is given by the antenna's *gain* in that direction. This property is dependent on the size and shape of the antenna structure and on the frequency of radiation as well. We will be dealing mainly with parabolic dish antennas in this chapter, and a simple relationship for gain in the boresight direction for such an antenna is given by equation 5-5.

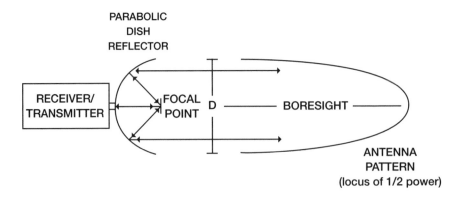

Figure 5-4. Parabolic antenna. Parabolic antennas are used to concentrate transmitting or receiving energy in a particular direction.

$$G_{dish} = \frac{4\pi A_e}{\lambda^2} \qquad (5\text{-}5)$$

where λ is the wavelength of the signal and **Ae** represents the *effective antenna area* which, for a parabolic dish, is given by equation 5-6.

$$A_e = \mu\left(\frac{\pi D^2}{4}\right) \qquad (5\text{-}6)$$

where D is the dish diameter (as depicted in Fig. 5-4) and μ is an *antenna efficiency factor* (typically around 0.5–0.8). Notice that using the same units for A_e and λ results in a dimensionless value for the gain.

As you can imagine, directions away from the boresight of a parabolic dish antenna reduce the effective area that can be seen, decreasing the associated gain of the antenna in that direction. The antenna pattern shown for the parabolic dish in the sketch also represents the antenna's gain. Though it is easy to imagine this as similar to a flashlight beam emanating from the antenna (which it is), it must be realized that this same gain applies to the antenna when it is receiving signals as well as when transmitting them. Parabolic antennas send and receive signals best from the boresight direction and do not receive signals as well from other directions. Numerically, the same relationship that gives the boresight gain for a transmitting dish (G_T), also applies to the antenna's ability to receive signals from the boresight direction, called the receiving gain (G_R).

Power Budget

The amount of power that arrives at an antenna, and the ability of the antenna and receiver to pick up and recognize that signal, determines if successful communications will occur. This is determined by a *link* (or *power*) *budget* analysis of the specific system, as depicted in Figure 5-5.

A certain amount of transmitter power (**P_T,** usually described in watts [W] or decibel-watts [dBW]) is delivered to the antenna. If this were an isotropic radiator, the power would be radiated uniformly in all directions and at some distance (**R**) from the antenna the amount of power detected would be:

$$\frac{P_T}{4\pi R^2}(W/m^2)$$

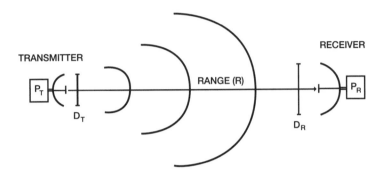

Figure 5-5. Link budget. A certain minimum amount of power must be received for successful communications.

because the power would be spread out throughout the 4π radians of the sphere around the isotropic radiator. Notice that this is an *inverse square relationship* which describes the loss of power of *any* electromagnetic signal propagating through space.

The gain of the antenna concentrates the power in the boresight direction and represents a multiplicative factor for the power in that direction. At a distance of **R** along the boresight direction from a parabolic antenna, the power would be:

$$\frac{P_T G_T}{4\pi R^2} (W/m^2)$$

The above relationship gives the *power per unit area* at range **R**. The amount of that power that falls on the receiving antenna depends on that antenna's effective antenna area A_e, the expression for which was given in equation 5-6.

$$\frac{P_T G_T A_e}{4\pi R^2} (W)$$

Relating this result to the form of equation 5-5 for the receiving antenna gain (G_R), we can write:

$$\frac{P_T G_T G_R \lambda^2}{(4\pi R)^2} (W)$$

We can define a term known as the *free space loss* (**L$_S$**) which represents the losses due to the spreading out of the signal through space between the transmitter and receiver:

$$L_S = \left(\frac{4\pi R}{\lambda} \right)^2 \qquad (5\text{-}7)$$

Incorporating the free space loss term, our equation simplifies to:

$$\frac{P_T G_T G_R}{L_S} (W)$$

Finally, this expression usually incorporates a term that takes into account a multitude of other signal power losses due to factors such as the atmosphere, ionosphere, electromagnetic interference, and the like. This term is called the *incidental loss* term (L_I) which, as a loss term, appears in the denominator of the above equation. The final form of this equation is called the *link* (or *power*) *budget* which gives a value for the power received by a station at a distance **R** from a transmitter:

$$P_R = \frac{P_T G_T G_R}{L_S L_I} (W) \qquad (5\text{-}8)$$

In many cases this calculation is made using decibel values which would, for the above relationship, be written as:

$$P_R = P_T + G_T + G_R - L_S - L_I \,(dBW) \qquad (5\text{-}9)$$

where each term is either specified in or converted to dB using equation 5-3.

Noise. The amount of power received may still not determine if the receiver will understand the transmitted information, which is required for successful communications. The reason for this is the presence of noise which can interfere with the transmitted signal. There are many sources for this noise including the transmitting and receiving equipment as well as natural causes such as the sun, earth, atmosphere, and even other celestial bodies. The amount of noise a source produces is represented by the source's *equivalent noise temperature* (T_{eq}). For our purposes, we will assume that the receiving equipment is the major source of noise and that its noise temperature will be known or given. The resulting noise is given by:

$$N_O = kT_{eq} \, (bw) \, (W) \tag{5-10}$$

where k = Boltzmann's constant ($k = 1.38 \times 10^{-23}$ W/°K $\cdot$ Hz) and **bw** represents the *bandwidth* or the range of frequencies the receiver has been designed to pick up. Less noise may be experienced by using a small operating bandwidth, but in all practical cases the desired information signal is spread out over a range of frequencies dictating what the minimum bandwidth for a receiver must be.

Signal-to-Noise Ratio. Successful communication systems (known as *links*) are designed to operate under all expected conditions of losses and noise. The measure of this is given by a receiver's *signal-to-noise ratio*, which is given by equation 5-11:

$$S/N = \frac{P_R}{N_O} \tag{5-11}$$

A receiver will have a *minimum* signal-to-noise ratio specified for acceptable operation. If the received power decreases or noise increases such that the actual signal-to-noise ratio is less than the minimum specified for a receiver, the information in the carrier signal may not be intelligible or the carrier signal may not be detected at all.

Example Problem:

A geostationary communications satellite has a traveling-wave-tube (TWT) amplifier with a maximum power output (P_T) of 9 W and transmits via a 5 m diameter parabolic dish antenna with an efficiency of 0.6. The antenna is boresighted on a ground station using a 30 m diameter parabolic dish antenna ($\mu = 0.5$). The receiver bandwidth is 1 MHz wide around the 6 GHz down-link frequency, and the receiving components are cooled to maintain an equivalent temperature of 300 °K. Slant range (distance between antennae) is 40,000 km and the incidental down-link power losses (L_I) are 1 dB. Determine the signal-to-noise ratio of the receiver.

Solution:

$$
\begin{aligned}
Ae_{(T)} &= 11.78 \text{ m}^2 & &= 10.71 \text{ dB} \\
Ae_{(R)} &= 353.43 \text{ m}^2 & &= 25.48 \text{ dB} \\
G_T{}^{(1)} &= 59{,}217.63 & &= 47.73 \text{ dB} \\
G_R{}^{(1)} &= 1{,}776{,}528.8 & &= 62.5 \text{ dB} \\
L_S{}^{(2)} &= 1.011 \times 10^{20} & &= 200.05 \text{ dB} \\
P_R{}^{(3)} &= 7.44 \times 10^{-9} \text{ W} & &= -81.3 \text{ dBW} \\
N_o &= 4.14 \times 10^{-15} \text{ W} & &= -143.83 \text{ dBW} \\
P_R/N_o &= 1.8 \times 10^6 & &= 62.55 \text{ dB}
\end{aligned}
$$

Notes: (1) $\lambda = c/f = 0.05$ m
     (2) R and λ must use same units
     (3) 1 dB = 1.26 decimal

MODULATION

Modulation is the means by which information is impressed on electromagnetic radiation so that it may be transmitted from one location to another. There are at least two different components to a communications signal: the carrier wave and the baseband signal.

The *baseband signal* represents the information desired to be transmitted. In the case of a radio station, this may be the audio frequencies (of a few thousand hertz) of the DJ's voice or the music of a record. These are good examples of *analog* signals which are the raw, continuous wave-like frequencies of sound or light. Television stations transmit higher baseband frequencies that are used by a TV set to illuminate the tube in a certain way to produce a picture. Banks and businesses (and satellites) may convert their baseband information (numerical transactions, written text and pictures, remotely sensed frequencies) into *digital* signals (strings of ones and zeros) which then *represent* the original information which may allow improved transmission of the data.

Carrier waves, as the name implies, are electromagnetic signals used to carry the baseband information between the transmitter and receiver. Carrier waves are used because the baseband signal may not have sufficient propagation characteristics for successful communications. For instance, audio frequencies are attenuated severely by the atmosphere, limiting their useful ranges (can you imagine radio stations that simply blasted their music from loudspeakers all over the city?). The particular carrier wave frequency may be chosen primarily for its propagation characteristics. For example, if communications with a satellite in orbit is desired, the signal frequency must be above the critical frequency of the ionosphere or the signal will not pass through into space.

As important as its propagation characteristics, carrier waves are also chosen according to the amount of baseband information that must be transmitted. Only so much information can be impressed upon a carrier wave and still be recovered when the signal is received (and demodulated) by the receiver. As a rough rule of thumb, it can be considered that *a carrier wave can only be impressed with baseband frequencies up to about 10% of the carrier wave frequency.* So, to transmit the human voice (around 4 kHz) a minimum carrier frequency of about 40 kHz would be required.

As mentioned above, the baseband information must be impressed onto the carrier wave for propagation between stations. In all cases, some characteristic of the carrier wave, such as amplitude or frequency is modified to represent the baseband information whether it is in analog or digital form. The following sections describe the general modulation techniques using analog baseband signals. Digital communication techniques are very similar to these analog techniques and will be presented later in this chapter after digital signals themselves are more fully described.

Amplitude Modulation (AM)

In amplitude modulation schemes, the baseband signal is used to change the amplitude of the carrier wave in such a way as to represent the baseband information. Figure 5-6 shows how an analog baseband signal modifies a carrier wave to produce an AM signal.

Notice that the frequency of the carrier wave has not changed. Only the amplitude varies and it does so in exactly the same manner as the baseband signal. The AM signal will propagate between stations with the characteristics of the carrier wave, but if we only look at its amplitude, we can see the characteristics of the information impressed upon the signal.

Frequency Modulation (FM)

In frequency modulation, the frequency of the carrier signal is modified by the baseband signal, as illustrated in Figure 5-7. Note that the amplitude of the signal remains constant. The frequency variation from the known carrier frequency is directly proportional to the frequencies of the baseband (information) signal.

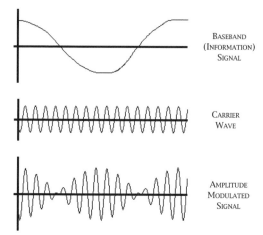

Figure 5-6. Amplitude modulation. The amplitude of the carrier wave is changed in a way to represent the information to be transmitted.

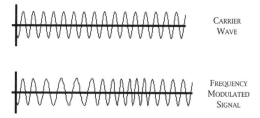

Figure 5-7. Frequency modulation. The frequency of the carrier wave is changed in a way to represent the information to be transmitted.

Phase Modulation (PM)

Information can also be impressed upon a carrier signal by using the baseband signal to modify the phase of the carrier wave, called phase modulation. The rate at which the carrier phase is changed, and/or the location along the wave at which the change occurs, can be used to represent the baseband signal.

An example of a PM signal is shown in Figure 5-8. Notice, in this case, that both the amplitude and frequency of the signal remain constant and that only the phase changes, as indicated by the arrows in the figure. Phase

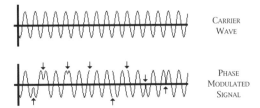

Figure 5-8. Phase modulation. The phase of the carrier wave is changed in a way to represent the information to be transmitted.

modulation is not used much in conventional (analog) communications systems, but is widely used in digital communications schemes employed by many satellite systems.

Demodulation

Demodulation is defined as the recovering of the baseband signal from the received modulated signal. The basic process used is simply the opposite of that used to modulate the carrier signal in the first place. For AM signals, a device called an envelope detector senses the change in amplitude of the received signal and reconstructs the baseband signal from this. Similar methods are used by FM demodulators (frequency detectors) and PM systems (phase detectors). It is very important that the receiver be able to reproduce the carrier wave exactly, as this is used by the demodulator to "subtract" this component from the received signal (when you "tune" in a station you are adjusting the receiver's oscillator to match the desired carrier frequency).

Though demodulation may sound simple, the procedure is complicated by the fact that the received signal is usually very weak, with the information-carrying portion of the signal (P_R) possibly weaker than the noise signal (N_o). Also, due to its transmission through the atmosphere (and perhaps the ionosphere), the signal may be significantly corrupted (changed by interaction with the elements) which may affect proper recovery of the information signal.

DIGITAL COMMUNICATIONS

Digital communications offers some benefits over analog communications, mainly in terms of the transmission, reception, and handling techniques of the digitized information. Since more and more systems are incorporating solid-state technology (which is itself a digital medium), the use of digital signal processing has increased as well. Some familiar examples are CDs (compact disks), on which music is stored digitally and reconverted into analog sounds by your player, and telephone links (fiber-optic digital telecommunications). Even radio and television stations are beginning to digitize their signals. In this section, we will discuss how a baseband signal (information) is digitized, what the characteristics of such signals are, and how the resulting signal is modulated and demodulated.

Basic Binary

Digital signals are merely *representations* of the original information using a simple (binary) coding system. The invention of the transistor in the 1950s began the interest in this field. With the transistor, it was possible to represent a situation ("true" or "false," for instance) by indicating either "true" by allowing current flow through the device (device "on"), or "false" with no current flow (device "off"). Since only two situations are possible, a transistor is a *binary* (2-ary) device, and in binary language, "on" could represent a digital "1" and "off" a digital "0."

A single transistor can only represent two situations, or *states*, at any one time (either **1** *or* **0** could be indicated). If two transistors are combined simultaneously, four different states could be represented at any one time (**0** and **0**, **0** and **1**, **1** and **0**, *or* **1** and **1**, respectively, on the two transistors). Three transistors allow eight different states to be represented, four transistors 16 states, and so on. Table 5-1 shows the possible states combinations of three and four transistors. (Note: Each **1** or **0** that represents a possible state for a transistor is known as a bit.)

Notice that, in going down the table, to get from one state to the next on the list you need only add (in binary) a "one" to the preceding value. There is an entire branch of mathematics concerned with the manipulation of binary numbers, but for our purposes we need only the simple relationships which relate the number of states that a certain number of bits can represent:

$$2^n = k \qquad (5\text{-}12)$$

and

$$n = 1.4427\ln(k) \qquad (5\text{-}13)$$

Table 5-1
Number of Bits vs. Number of States

3 BITS			4 BITS			
0	0	0	0	0	0	0
0	0	1	0	0	0	1
0	1	0	0	0	1	0
0	1	1	0	0	1	1
1	0	0	0	1	0	0
1	0	1	0	1	0	1
1	1	0	0	1	1	0
1	1	1	0	1	1	1
8 STATES			1	0	0	0
			1	0	0	1
			1	0	1	0
			1	0	1	1
			1	1	0	0
			1	1	0	1
			1	1	1	0
			1	1	1	1
			16 STATES			

where **n** is the number of bits that can represent **k** different states. For instance, a typical personal computer has a single chip processor with a 16-bit capability allowing 65,536 different states. This means the computer can perform 65,536 different operations, with each letter, number, symbol, or function counting as, at least, one operation. This really is quite a capability!

As mentioned above, each **1** or **0** is known as a *bit.* In computer lingo, four bits put together are called a *nybble,* and eight bits constitute a *byte.* A *word* is a combination of bits (or nybbles or bytes) used to represent something of value to the user. Of course, everyone using the same information must know how many bits make up each word, as well as how the bits are arranged or ordered, or else the information would just be a meaningless stream of ones and zeros.

Analog to Digital (A/D) Conversion

The following sections discuss how an analog signal (like music, a voice, or even visible frequencies) is converted into groups of digital

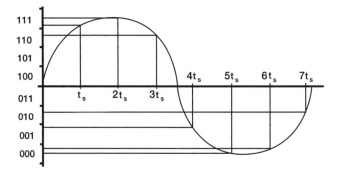

Figure 5-9. A/D conversion. An analog signal may be sampled periodically and given digital values used to represent the original signal.

words that can represent the original signal. Figure 5-9, which will be referred to in the following discussions, depicts a single sinusoid of an analog signal being digitized.

Quantization. Quantization is the method of assigning a *range* of digital values to represent the original signal. Two factors limit this process: the range of *signal* values that must be covered, such as amplitude range or frequency range, and the number of bits that the system can use to represent these values. For the signal in Figure 5-9, the system must be able to represent the signal from peak to subsequent peak of the sinusoid shown. With only a single bit we could indicate that the sinusoid was either above (with a "one") or below (with a "zero") the zero amplitude line, but nothing more. With two bits we could split the above and below portions of the sinusoid in half and could indicate (still crudely) how far above or below the signal was at that point. The signal shown in the figure has been split into eight quantization levels (known as *Q-levels*) giving a better ability to represent the signal amplitude. As you can imagine, with an unlimited number of levels we could represent *exactly* where above or below the line the signal was at any time. Unfortunately, there are limitations to the number of bits that a system can use (as we will discuss shortly) and we find that we must try to represent the signal with as few bits as possible.

We can use equations 5-12 and 5-13, where **k** now represents the number of Q-levels and **n** the number of bits, to determine the number of bits required or the number of levels available given one of the parameters as a system requirement.

Coding. The method of assigning a particular bit word to each quantization level is called *coding*. (Coding for security purposes is another separate process, the mechanics of which are obviously more complicated than the simple digitization coding discussed here.) Because the signal in the figure has been split into eight Q-levels, three bits are required to describe each level separately. The coding used is the *specific order* in which the eight different three-bit words are assigned to each level. For the signal shown in Figure 5-9, **000** has been assigned to the bottom level and a binary "one" is added to get the coding for each subsequent level.

Sampling. Quantization values are assigned to a signal often enough to allow the receiver of the information to reconstruct the original signal (sort of by simply connecting the dots indicated by the digital information). The frequency at which a signal is assigned quantization levels (sampled) is called the *sample period* (t_s).

It should be obvious that if a signal were sampled too infrequently, not enough information would be available to reconstruct the original signal. However, as with quantization, sampling too often may result in too large a number of bits for the system to be practical. It has been shown that if a signal is sampled at least twice during each sinusoid, there will be enough information to reconstruct the signal. In terms of frequencies:

$$f_s => 2f_{max} \tag{5-14}$$

where f_s represents the *sampling frequency* (samples/sec), and f_{max} represents the *highest* frequency in the range of signals being sampled.

Note that the sample period is simply the inverse of the sampling frequency as shown in equation 5-15.

$$t_s = \frac{1}{f_s} \tag{5-15}$$

Transmission Rate. The transmission rate represents the number of bits per unit time that a digital system creates. Since a sample of **n** bits is being taken every t_s seconds, and since $f_s = 1 / t_s$, the transmission rate (TR) can be found from:

$$TR = \frac{n}{f_s} = n \times f_s \text{ (bps)} \tag{5-16}$$

where **bps** stands for bits per second. The transmission rate is an important parameter of a digital communication system, as it is a measure of the capabilities of the system to transmit information. Most modern spacecraft use digital communication methods, and in many cases the transmission rate is a major factor limiting the system capability. Examples are telecommunications satellites which can only handle so many telephone calls at a time or remote sensing systems that can only observe so much information due to data storage and transmitting limitations. The major limiting factor to transmission rate (and subsequently, the limit to the sample rate and number of Q-levels) is the carrier wave's ability to transport digital information, as described in the next few sections.

Example Problem:

Satellite transmission of telephone calls allots a bandwidth (range of frequencies) of 4 kHz (associated with the minimum bandwidth required to reproduce the human voice) for each call. If eight Q-levels are used to digitize each call, determine the minimum sampling rate (frequency), sampling period, and transmission rate associated with one conversation.

Solution:

$$f_{s_{min}} = 2\,f_{max} \quad = 8{,}000 \text{ samples/sec}$$
$$t_s = 1/f_s \quad = 0.125 \times 10^{-3} \text{ sec}$$
$$n = 1.4427\,\ln(k) = 3 \text{ bits/sample}$$
$$TR = n \times f_s = n/t_s = 24{,}000 \text{ bits/sec (bps)}$$
$$= 24 \text{ Kbps}$$

Digital Baseband Signals

The output of an A/D converter is a *string* of "one" and "zero" values representing the bits assigned to each successive sample of the original signal. These values are usually output as voltages, where a positive voltage may correspond to a "one" and either a negative or zero voltage would represent a "zero" bit. Figure 5-10 shows these two common methods of representing a digital string of values.

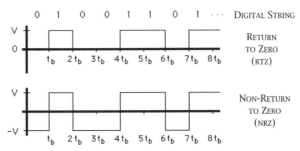

Figure 5-10. Digital baseband signals. The "ones" and "zeros" of digital baseband signals are usually represented by voltage levels held for a specified period of time.

The time that the specific voltage is held to represent each bit is called the *bit period* (t_b), as indicated in the figure. It is desirable to represent each bit as long as possible, as this would allow a receiver more time to recognize each bit as it is received. But for real-time (continuous) communications, the number of bits representing each sample must be transmitted before the next sample bit stream comes up. To ensure this, the bit period must be less than or equal to the sample period divided by the number of bits per sample:

$$t_b = \leq \frac{t_s}{n} \; (\text{sec/bit}) \tag{5-17}$$

The pattern of this voltage output now represents the information to be transmitted and will be used as the baseband signal to modulate the carrier wave in the methods described next.

Digital Modulation Methods

The digital baseband signals that are the product of A/D conversion are used in methods similar to those discussed earlier in analog modulation schemes to impress information onto a carrier wave for transmission between stations. It was mentioned earlier that analog baseband signals were limited in frequency to about 10% of the carrier wave, and we find that digital signals have similar limits on the number of bits that a carrier wave can transport. We can simply consider a digital baseband signal as a continuous waveform with a period equal to two bit periods, as depicted in Figure 5-11 which shows the first few bits of a NRZ waveform.

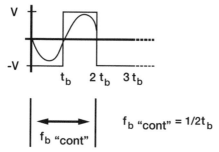

Figure 5-11. Digital/continuous signal analogy. To determine the information-carrying capability of digital modulation techniques, digital signals can be likened to an equivalent analog signal.

The maximum transmission rate a digital system could have, based on a given carrier wave frequency (f_c), can be found from the following relationships:

$$TR_{max} = \frac{1}{t_b} = 2 \times f_{b\text{"cont"}} = \frac{f_c}{10} \tag{5-18}$$

where $f_{b\text{"cont"}} = \frac{1}{2} t_b$, representing the equivalent "continuous" frequency of the digital waveform. Once a maximum transmission rate is determined, the bit period, sampling frequency, and sample period can be established.

Example Problem:

A communications satellite uses a 6 GHz down-link frequency to transmit digitized telephone calls. Using the results of the last example problem, determine the maximum transmission rate and the number of calls that may be transmitted simultaneously.

Solution:

From the last example problem we found that for each call:

$f_{s_{min}} = 8,000$ samples/sec
$t_s = 0.125 \times 10^{-3}$ sec
$n = 3$ bits/sample
$TR = 24$ Kbps

The *maximum* transmission rate for the satellite is:

$$TR_{max} = f_c/10 = 6 \times 10^8 \text{ bps} = 600 \text{ Mbps}$$

which indicates that simultaneously the link could handle:

$$TR_{max}/TR \text{ (one call)} = 25,000 \text{ calls}$$

assuming that the baseband information for each call can be combined into a single continuous bit stream. The methods used to accomplish this are beyond the scope of this text, but notice that the maximum period for each bit on the carrier wave must be much less than that for each call separately:

$$t_{b \text{ (one call)}} = t_s/n = 4.2 \times 10^{-5} \text{ sec/bit}$$
$$t_{b \text{ (link)}} = \tfrac{1}{2} f_{b\text{"cont"}} = 1/TR_{max} = 1.7 \times 10^{-9} \text{ sec/bit}$$

This result reveals that less than two nanoseconds can be allotted for each bit transmitted!

The following sections describe the basic digital modulation methods and their similarity to the analog modulation techniques discussed earlier.

Amplitude Shift Keying (ASK). In this modulation method, the carrier wave is either transmitted (representing a digital "one") or not transmitted (signifying a digital "zero") each bit period. The relationships between the

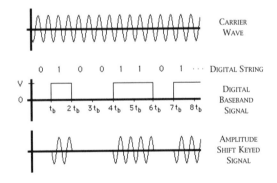

Figure 5-12. Amplitude shift keying modulation. The amplitude of the carrier wave is used to represent the digital baseband signal.

baseband signal, carrier wave, and the resulting ASK signal are shown in Figure 5-12.

Frequency Shift Keying (FSK). In this modulation scheme, two carrier frequencies are used, one to represent a transmitted "one" and the other a digital "zero." An example of an FSK signal is shown in Figure 5-13.

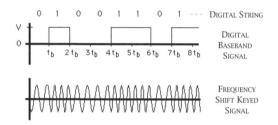

Figure 5-13. Frequency shift keying modulation. The change in frequency of the carrier wave is used to represent the digital baseband signal.

Phase Shift Keying (PSK). In phase shift keying, a change in a bit state (from "zero" to "one" *or* from "one" to "zero") is indicated by a change in the phase of the carrier wave. Notice that if there is no change in the bit pattern, the phase of the carrier signal doesn't change until there is one, as shown in Figure 5-14.

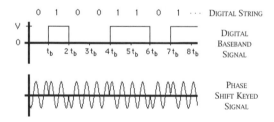

Figure 5-14. Phase shift keying modulation. The change in phase of the carrier wave is used to represent the digital baseband signal.

An important point in PSK systems is ensuring that the receiver knows what the starting bit is supposed to be, or else the reconstructed digital string will be opposite that transmitted. This is usually done by first sending a code which, whether received correctly or not, will indicate what the first bit in the message should be.

Demodulation. Demodulation methods are similar to those described earlier for analog signals in their use of envelope, frequency, and phase detectors. For digital demodulation, the demodulator must only detect the presence of a transmitted "one" or "zero" which is relatively more simple than recovering the actual analog baseband signal from the received signal (plus noise). This is one of the advantages of digital communications over analog systems, as it allows a successful link with lower minimum signal-to-noise ratios.

An important factor for successful use of digital signals for communications is ensuring that everyone on the link is synchronized in *time*. Modification of the carrier wave by the transmitter modulator only occurs each bit time (which, as we've seen, can be measured in nanoseconds), and digital demodulators are looking to detect changes in the received signal at these times only as well.

COMMUNICATIONS SYSTEMS

Figure 5-15 displays the components of a typical communication system. In the transmitter, the *source* represents the originator of the information signal ($F_{[t]}$). This may be the microphone of a radio or telephone; the video receivers of a television camera; the antenna of a remote sensor; or whatever is picking up or originating the desired information, turning it into electromagnetic signals suitable for transmission. If the signal is to be digitized, an A/D converter would be part of the *preparatory electronics* that make a signal ready for modulation onto the carrier wave which occurs within the *modulator* section of the transmitter. The *amplifier*

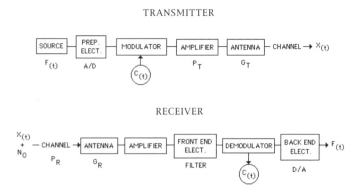

Figure 5-15. Communications systems. Many of the common components of a typical communications system are depicted.

boosts the signal up to the desired transmitter power (P_T) and delivers this to the *antenna* from which the electromagnetic signal ($X_{[t]}$) will propagate. The *channel* represents the medium through which the signal travels between stations. (For a radio station, this medium would be the atmosphere between the station and your car. The channel for telephone or cable TV is the line coming in your house.)

On the receiving end, the antenna picks up the signal (with its associated noise) from the channel. Since the power level received is usually quite low compared to that which was transmitted, the signal is delivered directly to an *amplifier* to boost the signal strength. The *front-end electronics* attempt to filter out some of the signal noise before delivering the signal to the *demodulator,* which recovers the baseband signal from the carrier wave frequency. (Note: As was mentioned earlier, the receiver electronics contribute significantly to the overall equivalent noise temperature of a communications link. To minimize this contribution, many receiving stations cool their front-end electronics, some down to near absolute zero using liquid nitrogen, to improve the received signal-to-noise ratio). Finally, the *back-end electronics* reproduce the transmitted information in its original form (whether by speaker, TV tube, teletype, etc.). It is here that the digital-to-analog (D/A) converter will reconstruct the original signal if it was first digitized before transmission.

REFERENCES/ADDITIONAL READING

Pritchard, W., Suyderhoud, H., and Nelson, R., *Satellite Communication Systems Engineering,* 2nd ed. Englewood Cliffs: Prentice-Hall, Inc., 1993.

Feher, K., *Digital Communications, Satellite/Earth Station Engineering.* Englewood Cliffs: Prentice-Hall, Inc., 1983.

Roden, M., *Analog and Digital Communication Systems.* Englewood Cliffs: Prentice-Hall, Inc., 1979.

Gagliardi, R., *Introduction to Communications Engineering.* New York: John Wiley & Sons, 1978.

EXERCISES

1. A ship at sea wishes to communicate with a shore station over the horizon using its HF radio. Using the F_2 region of the ionosphere for a "reflection" point between the stations, it is determined that the inci-

dence angle must be 30° to get the range required. Determine the highest frequency that the ship can use to transmit its information. Would this be a sufficient carrier frequency to transmit a 3 MHz baseband signal? (Explain.)

2. A geostationary communications satellite has a traveling-wave-tube (TWT) amplifier with a maximum power output (P_T) of 9 W and transmits via a 5 m diameter parabolic dish antenna with an efficiency of 0.6. The antenna is boresighted on a ground station with a 30 m diameter parabolic dish antenna (efficiency = 0.5) with a receiver cooled to maintain an equivalent temperature of 300° K. Slant range (distance between antennae) is 40,000 km, and the down-link frequency is 6 GHz. The incidental down-link power losses (L_I) are 1 dB.

   Do this problem twice, once using the decimal form of the equations and values, and then again using dB values to perform the calculations.

   a. Compute the gains of the transmitting antenna (G_T) and the receiving antenna (G_R).

   b. Compute the free-space loss (L_S).

   c. Determine the power received by the ground station (P_R).

   d. What is the received signal-to-noise ratio (P_R/N_o)?

   e. Show that the values obtained in part (d) above, using the decimal approach and the dB approach, are equal.

3. *Voyager 2,* sent out to visit the outer planets, is in a hyperbolic trajectory which will carry it out of the solar system. Scientists are interested in monitoring *Voyager 2* for as long as possible to see if there is an interface between the space in the vicinity of our sun and true interstellar space (the heliopause), and what the interstellar space may be like. *Voyager 2* has a 3.7-meter diameter parabolic dish antenna with an efficiency factor (μ) of 0.61. The onboard 21.3 W transmitter sends data back towards the earth on a down-link frequency of 8.415 GHz (X-band). Signals from *Voyager 2* are picked up by NASA's Deep Space Network (DSN) parabolic antennas that are 64-meters in diameter (μ = 0.42) with receivers that are cooled down to 28.5° K. Incidental losses are estimated to be 1.549 dB and the minimum signal-to-noise ratio that the DSN receivers can tolerate is 2.05×10^5.

   a. Determine the signal power (in watts) received from the spacecraft by the DSN at the time of minimum S/N reception.

   b. Compute the distance from the earth at which contact will be lost (in terms of A.U.s where 1 A.U. = 149.5×10^6 km).

c. Will we still be in contact with the spacecraft when it flies beyond the *farthest* reaches of Pluto? ($a_{Pluto} = 5.9 \times 10^9$ km, $e_{Pluto} = 0.25$)

4. An FM radio station wishes to transmit baseband frequencies from 20 to 20,000 Hz in order to broadcast good quality music. The station wants to transmit digitally (by first converting the baseband signal to a digital signal) using *byte*-sized digital words.

    a. Determine the number of Q-levels the signal may be assigned values from.

    b. What is the minimum sampling frequency (and associated sampling period) that may be used?

    c. Determine the bit period, the transmission rate, and the minimum carrier frequency that might be used to transmit the music.

    d. Can the current frequency range assigned to FM broadcasting be used by this station? (Explain.)

5. The Army and Marine Corps would like to have the capability for satellite communications with ground forces using a portable transceiver. Such a radio would have a small (1 meter) diameter parabolic dish antenna with a low efficiency (0.5). Maximum transmitted power would be 2 watts. The soldier would like to be able to transmit to a geostationary satellite down to 5 degrees elevation angle ($\psi = 5°$ [described in Chapter 6], $R_{slant-max} = 41,126.7$ km). The satellite has a 5-meter diameter parabolic dish (efficiency = 0.9) and uses deep space as a heat sink to keep the receiver cooled to 75°K. Up-link frequency is 30 GHz and worst case ($\psi = 5°$) incidental losses are 4 dB. Consider the antennae as being able to boresight on each other during communications, and determine the minimum signal-to-noise ratio the satellite will have to be able to handle for a successful link. While showing your calculations for the link budget, put a box around the following parameters:

A_{e_T}, A_{e_R}, G_T, G_R, L_S, P_R, and N_o

CHAPTER 6

Remote Sensing

On August 7, 1959, *Explorer 6* sent back the first satellite-produced picture of the earth taken from an altitude of about 20,000 km in space. Although grainy and lacking detail, it gave a spectacular view of swirling cloud formations over the oceans and landmasses of a large portion of the earth. Perceiving the immense value of observations from space, planners quickly began to take advantage of the benefits of this perspective view, and a few months later, on April 1, 1960, *TIROS 1* was launched. *TIROS 1* ushered in an age of continuous operational meteorological satellites (often referred to as weather satellites) by returning thousands of pictures of the earth during its two-and-a-half-month lifetime. These and many other pioneering satellites demonstrated the basic advantages of satellite remote sensing: the relatively large (compared to airborne perspectives) amount of the earth in view at any time; the rate of coverage possible from a space-based platform; and the ability to view even the remotest locations on the earth's surface. For example, a typical polar orbiting weather satellite at an altitude of 830 km circles the earth every 102 minutes, moving with an earth velocity of nearly 6,600 m/sec, viewing thousands of square kilometers each second as its sensors scan cross-track from horizon to horizon. Two of these operational meteorological satellites obtain a global view of the earth twice a day, providing government and private weather services with surface and atmospheric data that improve and extend weather prediction. A stark contrast and testimony to the value of satellite weather observation is the accuracy and timeliness of observations of severe storms, such as hurricanes, compared to presatellite times when we relied on data from ships of opportunity or sparse information from weather aircraft designated to search for hurricanes. Although property damage still occurs, the advanced warning provided by timely space-based weather observation mitigates damage and has sharply reduced loss of life.

There are a multitude of remote sensing satellites that enjoy the coverage benefits offered by the vantage point of space. In addition to environmental monitoring satellites, there are a host of fine spatial resolution satellites that owe their heritage to the pioneering work done by NASA's Landsat and Seasat satellites. These satellites employ electronic cameras and special radars to observe the earth and its oceans in great detail for military, scientific, and commercial purposes.

Remote sensing from space can be categorized in a variety of different ways, such as the type of sensor employed, the application purpose, and/or the sponsorship of the mission. These themes will be expanded in the sections that follow but are first outlined in this introduction.

Sensors

Sensors can first be categorized as being passive or active. A passive sensor receives information from the earth and the atmosphere. A camera is a good example of a passive sensor relying on solar illumination of the object(s) in view to provide input to the camera. By contrast, an active sensor provides the source of energy used to view the area of interest. Radars and lasers are examples of active sensors.

Both passive and active sensors operate over a wide portion of the electromagnetic spectrum. The precise frequencies or wavelengths chosen are a function of first, the application, and second, the influence of the propagation path between the satellite and the earth. The signatures provided by the earth and its atmosphere result from highly complex interactions of induced, in the case of an active sensor, or natural, in the case of a passive sensor, electromagnetic radiation. These interaction mechanisms include reflectance, emission, absorption, and scattering, and the proper blend of sensor wavelengths can reveal information about the types and condition of soil, vegetation, water, ice, and the atmosphere. One of the most popular regions of the electromagnetic spectrum employed by remote sensors are the visible bands that provide picture-like observations of the earth's surface and the clouds that envelope the earth. However, many other portions of the electromagnetic spectrum are used to provide additional information about the earth and its resources. Infrared and microwave sensors provide information on atmospheric, surface, and even subsurface conditions. Microwave sensors have the added benefit of being able to observe the earth's surface through clouds and the capability of operating day or night.

The combined data from passive and active sensors, spanning the electromagnetic spectrum, allows users to study a host of important ecological, economic, and environmental subjects including crop coverage, health, and yield; land use and erosion; ocean and coastal conditions; and the effects of man on his environment in the form of pollution or depletion of the amount of ozone in the upper atmosphere.

Applications and Sponsorship

The genesis of remote sensing missions is usually associated with some applications goal that is sponsored by a civil governmental, military, or commercial organization. First efforts are usually exploratory research or proof-of-concept systems used to find the right sensor combinations and orbits for observation. If these research efforts prove successful, and a sustaining need can be established for the information that can be provided by the space-based remote sensing system, then a decision may be made to continue with a series of operational satellites that provide a continuum of information. For example, the aforementioned Tiros experiments led to the NOAA (National Oceanic and Atmospheric Administration), GOES (Geostationary Operational Environmental Satellite), and military DMSP (Defense Meteorological Satellite Program) satellite series. This evolutionary process has led to a number of established operational missions that continuously collect information about the earth, balanced by one-of-a-kind exploratory or scientific missions that constantly probe our earth, solar system, and universe in an effort to better explain the physical processes that govern our existence.

Any attempt to rigidly classify missions erodes under the progress of technology, budgets, and sovereign or national interests. At present, most developed nations have the capability to create, operate, and sustain a wide-ranging space remote sensing program that can be segmented into intelligence, military, civil governmental, and commercial operational programs as well as a varied set of progressive scientific exploration programs. With the exception of the protective and narrowly sovereign military and intelligence programs, many of the remaining mission categories are becoming shared efforts, with many nations combining their resources and technologies to produce cooperative benefits. The ever-expanding base of sensor technologies, missions, and international sponsors has produced a robust interest in, and reliance on, space remote sensing—all stemming from the fact that the vantage of space allows man to look at himself and his world in a global sense for the first time in history. In order to continue, it is necessary to describe the physics and processes employed by space remote sensing systems to gain more detailed information of interest to users on earth.

REMOTE SENSING PRINCIPLES

Remote sensing is the collection of information about an object without being in physical contact with the object. Space-based remote sensors are also used for reference and control functions such as providing input for spacecraft attitude control, weapons guidance, spacecraft planetary landing, and spacecraft docking. In the context of this chapter, remote sensing will be restricted to data collection and to methods that employ electromagnetic energy as the means of detecting and measuring target characteristics. This definition excludes electrical, magnetic, and gravity surveys that measure force fields rather than electromagnetic radiation.

Remote sensing systems employ the electromagnetic spectrum in unique ways to send and/or collect natural or man-induced signals. The use of particular wavelengths and signals for remote sensing can be based on:

- Natural radiation associated with the phenomena to be observed.
- Response of an object or material to an active signal created and transmitted to the target area by the remote sensor.
- The use of atmospheric or ionospheric interaction in the vertical path from the satellite to the earth, creating a class of sensors identified as sounders.
- The avoidance of atmospheric or ionospheric interference.

Earth viewing systems concentrate on the microwave, infrared, and visible regions of the spectrum from which information can be derived for a variety of specific imaging categories, including land mapping, specific target surveillance, and environmental monitoring. In addition to these more popular and commonly identifiable applications, space remote sensors yield important information on basic geophysical processes such as atmospheric temperature, water vapor, and liquid water; ocean surface wave height, length, and dynamics; surface wind speed and direction; surface temperature and ocean color; land area spectral evaluations; topographic variations; and object or geological classifications.

If the sensor relies primarily on an ability to discern the spatial characteristics of the objects in the field of interest, it is referred to as an *imager.*

Many sensors make primary use of the spectral characteristics of the response or the strength of the response to determine the context and content of the measurement and are referred to as *spectrometers* or *radiometers*. Whatever the primary method or measurement, all sensors have performance values described in terms of spectral (frequency-related), spatial (geometric), and radiometric (radiative-properties) dimensions. Moving or scanning sensors also provide a temporal (time-based) pattern of observation, sometimes referred to as the revisit interval. Current research and development activities lead the way to eventual space-based evaluation of ocean and land subsurface physical and biological characteristics including soil moisture and bathymetric measurements. Finer spectral and spatial observations of land areas are on the horizon, and even surface pressure monitoring from space appears to be possible in the future.

Classification of Remote Sensors

Earlier in this chapter, we identified passive remote sensors that collect electromagnetic energy. The camera (without lighting aids such as a flash unit) is the classic example of a *passive* sensor: Incoming light impinges on sensitive film which retains the image. *Radiometers* are devices that replace the film with detectors that are sensitive to the incoming radiation (which can be in other portions of the electromagnetic spectrum besides the visible frequencies) and convert the information to electronic signals that can be transmitted to and evaluated by users on the ground. Collection devices include lenses and antennas that focus the energy on detectors such as solid state materials, charge coupled devices (CCDs), and other spectrally unique and energy-sensitive conversion elements.

Sensors that create their own source of target or scene illumination are referred to as *active* sensors. These systems create and transmit energy that propagates from the satellite to the target, interacts with the target, and returns to the sensor. Active sensors include radars (altimeters, scatterometers, real aperture radar (RAR) imagers, synthetic aperture radar (SAR) imagers) and lasers (laser altimeters, scatterometers, and imaging devices). These instruments have the benefit of being able to control the time/frequency content, direction, and energy level of the illumination. Among other things, this characteristic permits active sensors to greatly improve on the spatial (geometric) performance of the system using methods which will be described later.

Geometry of Remote Sensing

The basic geometrical relationships between a satellite and the earth were described in Chapter 2. These are reproduced, along with some of the geometry specific to remote sensing, in Figure 6-1. The right side of the figure displays the angles described in Chapter 2 associated with the *angular field of view* (Ω) which represents the maximum angular distance visible from the satellite when the remote sensor's view is tangential to the earth's surface. The maximum *field of view* (FOV) can be described as the total curved earth area visible from horizon to horizon by an observation in space, associated with the swath width distance described in Chapter 2. In most cases, a remote sensor concentrates its instantaneous observation on a small portion of the total field of view using a lens, antenna, or some similar method of focusing incoming energy. This smaller area observed by the sensor at a particular time is known as the *instantaneous field of view* (IFOV) and will be described in more detail shortly.

The geometry associated with remote sensing is shown on the left side of Figure 6-1; these angles differ from those discussed in Chapter 2 in that they describe the relationship between the spacecraft and a point being viewed (the "target"). From the aspect of the satellite, the angle labeled θ_n is called the *look* (or *nadir*) *angle* and the complement of this angle is the *depression angle* (∂). The distance from the sensor to the target **R** is known as the *slant*

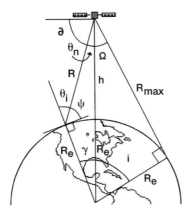

Figure 6-1. Remote sensing geometry. Some particular geometrical terms are used to describe the relationship between a remote sensor and the earth or its target.

range and is measured along the direction in which the sensor is looking (*boresight*). From the aspect of the point on the earth being viewed, the angle from the horizon to the satellite along the boresight is called the *elevation* (or *grazing*) *angle* (ψ), and the complement of this angle is called the *incidence angle* (θ_i). The angular distance from the nadir to the point being viewed, as measured from the center of the earth, is called the *earth angle* (γ).

Given the altitude of the satellite (**h**) and any one of the viewing angles, the slant range and all of the other angles can be calculated using:

$$R^2 = (h + R_e)^2 + R_e^2 - 2R_e (h + R_e) \cos(\gamma) \qquad (6\text{-}1)$$

and the law of sines:

$$\frac{R}{\sin(\gamma)} = \frac{R_e}{\sin(\theta_n)} = \frac{h + R_e}{\sin(\psi + 90)} \qquad (6\text{-}2)$$

Because the sensor may only be looking at a small portion of the available field of view at any particular time, a description of the area in view and a method of viewing the total area available must be developed.

Instantaneous Field of View. Just as a telescope of increasing power sees a smaller area on the face of the moon (for example), many remote sensors concentrate on a small portion of the FOV at any particular time. This smaller portion of the FOV visible by the sensor is known as the instantaneous field of view (IFOV). The remote sensor's ability to resolve spatial targets within the IFOV depends on the spatial acuity or dimensions of the detector's *picture element* (pixel) or *resolution cell* which is a characteristic of the physical make-up of the sensing device itself.

A passive remote sensor's IFOV is dependent upon the wavelength scaled dimension of the collection or focusing element (antenna or lens) and the distance from the remote sensor to the target area being observed, as shown by (equation 6-3), sometimes referred to as the *diffraction limit* of spatial performance.

$$IFOV = \left(\frac{\lambda}{D}\right) R \qquad (6\text{-}3)$$

where:
λ = primary wavelength of observation
D = dimension of the aperture (lens, antenna, mirror, etc.)
R = range from the sensor to the target area

In equation 6-3, the term λ/**D** is the sensor's *angular* IFOV expressed in radians. This angular field of view, projected at range **R,** produces an IFOV covering a *distance* equal to the angular IFOV multiplied by **R**. A two-dimensional aperture produces a two-dimensional pattern, as exemplified in Figure 6-2 which also shows the pattern projected onto the ground. The antenna used in the figure is not symmetrical and, hence, does not produce a symmetrical pattern on the earth. Notice the inverse relationship between the IFOV and the antenna dimensions, as expected from equation 6-3.

In the example shown, the antenna beam is pointed normal to the direction of satellite motion. Translation of the sensor IFOV onto the ground plane is known as the *ground instantaneous field of view* (GIFOV). Description of the GIFOV in the direction that the beam is pointing (referred to as the *range* or *cross-track* direction) involves simple trigonometry as shown in equation 6-4.

$$GIFOV_{RANGE} = \frac{\lambda R}{D_R \sin \psi} \qquad (6\text{-}4)$$

In the direction that the satellite platform is translating, referred to as the *azimuth* or *along-track* direction, the GIFOV is simply equal to the IFOV, as shown in equation 6-5.

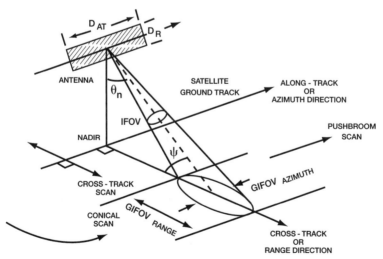

Figure 6-2. Instantaneous FOV and scanning. The field of view of the sensor can be superimposed upon the ground and scanned in several ways.

$$GIFOV_{AZIMUTH} = \frac{\lambda R}{D_{AT}}$$ (6-5)

Sensor Scanning. In most cases, a remote sensor would ideally cover as much of the observable area (field of view) as possible. To do this with an IFOV smaller than the FOV, the sensor must shift its concentration in some way, as a function of time, over the FOV. One important point to keep in mind is that the satellite (in all cases except for a geostationary satellite platform) is itself in motion over the planetary surface.

There are many methods that can be used to change the direction of the sensor's IFOV, including mechanically moving a telescope, mirror, or antenna; electronically shifting an antenna beam or optical path; and/or arranging a line or array of sensors with respect to the satellite's orbital motion. The scan can also follow a variety of patterns, some of which are illustrated in Figure 6-2. Many sensors scan from side to side across (normal to) the satellite nadir track. This type of pattern is known as a *cross-track scan*. Since the grazing angle and range are constantly changing as the scan gets farther from nadir, the GIFOV also constantly changes in the range direction.

Some sensors, such as microwave radiometers, hold a constant nadir (look) angle and then scan the sensor around the satellite nadir in a circular pattern known as a *conical scan*. Or a sensor can be held stationary and simply use the satellite motion to move the field of view across the surface in a technique called a *pushbroom scan*. Several modern optical sensors employ thousands of cross-track detectors in the sensor focal plane, and use the satellite's orbital translation to create such a scan. These and other variations on these themes are used to permit the selective IFOV of the sensor (representing the limit of a sensor's spatial acuity) to collect information over a broader range of the satellite field of view.

If a contiguous scan coverage is required, the scan rate (time for the sensor to complete one full sweep of the desired areas) is set so as to allow each scanned area to overlap slightly with previously scanned areas so that no spots are missed. Many sensors do not require a contiguous scan and instead may use a sampling scan to achieve their measurement goals.

Example Problem:

A remote sensing satellite is in a circular orbit at an altitude of 1,000 km. The sensor is a radiometer with its beam aligned normal to the satellite direction of flight and held at a fixed look angle of 30 degrees. This orientation is identical to the example shown in the previous figure. The aperture dimensions are 5 m ($D_{azimuth}$) × 1 m ($D_{cross-track}$) and the radiometer center frequency is 2 GHz. If the radiometer is held at a fixed look angle of 30°, determine the slant range to the ground, IFOV, and GIFOV for the along-track and cross-track dimensions.

Solution:

Using equation 6-2 (knowing the look angle, **h** and R_e) we can solve for all the other unknowns:

(ψ + 90) = 35.3° or *144.7°* (must be > 90°)

λ = 5.34°

R = 1186.8 km

From the center frequency of observation:

λ = c/f = 0.15 m

$IFOV_{along-track} = GIFOV_{along-track}$ = 35,610 m = 35.6 km

$IFOV_{cross-track}$ = 178.1 km

$GIFOV_{cross-track}$ = 218.2 km

Observation of Electromagnetic Radiations

The electromagnetic spectrum was described in Chapter 4 along with the relative frequencies and energies radiated by the sun and the earth due to their respective surface temperatures. In the case of passive sensors, these natural radiations, and their interaction with the earth and atmosphere, are what is sensed. Active sensors operate in exactly the same manner except that they provide a known spectral illumination of the target area and then sense the reaction of the target and atmosphere to the transmitted frequencies.

Target Spectral Interaction Mechanisms. A number of interactions are possible when electromagnetic energy encounters a target or scene. The interactions can take place in a plane or two-dimensional surface (*surface phenomena*) or in three dimensions, resulting in interactions called *volume phenomena*. These surface and volume interactions can produce a number of changes in the incident electromagnetic radiation, such as changes in magnitude, direction, wavelength, polarization, and/or phase. A remote sensor measures and records these changes, and the resulting data are interpreted to identify the characteristics of the target or scene that produced the changes in the electromagnetic radiation.

During the interaction between electromagnetic radiation and matter, both mass and energy are conserved according to basic physical principles. The following complex interactions may occur:

- Radiation may be *transmitted,* that is, passed through. The velocity of the electromagnetic wave may change as it is transmitted from a vacuum into other mediums. This velocity change is characterized by the medium's *index of refraction* (**n**) which relates the velocity of electromagnetic propagation in the medium (c_m) to that in a vacuum (**c**) as:

$$n = c/c_m \tag{6-6}$$

- Radiation may be *absorbed* and give up its energy, largely due to heating the substance or medium.
- Radiation may be *emitted* by elements in a scene as a function of element structure and temperature. As we saw in Chapter 4, all matter at temperatures above absolute zero emits energy.
- Radiation may be *scattered.* The scattering process may be preferential, tending to propagate in some specific directions, or it could scatter in all directions and be lost ultimately to absorption or further scattering outside of the sensor FOV. If the illumination source is the sensor itself, the energy returned to the sensor by this process is called *backscatter.*
- Radiation may be *reflected,* unchanged, from the surface of a substance with the angle of reflection equal and opposite to the angle of incidence.

These interactions with any particular target or form of matter are selective with regard to the wavelength of electromagnetic radiation and are specific for that form of matter, depending primarily upon its optical orientation with the sensor boresight, its wavelength scale-scattering surface or volume properties, and its atomic and molecular structure. These interactions between matter and energy provide the spectral and radiometric basis for remote sensing.

Radiative Transfer Equation. The amount of energy that a remote sensor receives from a target area is very much frequency-dependent and may come from a variety of sources as a result of the interactions described above. Figure 6-3 demonstrates the types of energies that a remote sensor may be looking to receive and reveals some of the interactions that energy may have during its transmission through the atmosphere.

In the figure, I_T represents the total energy received by the satellite sensor which has been designed to observe phenomena within a specific frequency range. In the situation depicted, the illumination of the scene is provided by the energy of the sun which has the known spectral characteristics described in Chapter 4. In other cases, the received energy may instead be produced by the natural radiations of the earth and atmosphere, or may be supplied by the transmitter of an active remote sensor. Each of the radiations which make up the total received energy is also a function of frequency and, for some observed frequencies, some of the components shown may not contribute to the received energy at all.

As shown in the figure, the sun's radiation, denoted by $\mathbf{F}_{(\lambda)}$, illuminates the scene from an angle θ_o called the *zenith angle*. One source of energy received by the remote sensor may come from some of this radiation that is scattered off small particles within the atmosphere. This *aerosol scattered* energy ($\mathbf{I_A}$) is uniformly radiated from the particles in all directions.

Another type of scattering of energy radiates in an anisotropic (not uniform) manner from the molecular constituents of the atmosphere. This

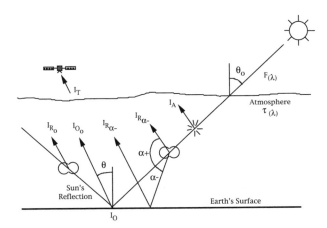

Figure 6-3. Remotely sensed energies. The energies received by a sensor in space may come from a number of sources and/or interactions.

Rayleigh scattering (pictured as a peanut-shaped pattern in the figure) is very direction-dependent and the scattering angles (denoted by α^+ and α^-) are a function of both the sun's zenith angle and the satellite look angle. The total amount of Rayleigh scattered energy received by the remote sensor may come from different sources: that scattered from direct illumination by the sun ($\mathbf{I_{R_{\alpha+}}}$), the reflection of this scattered energy off the surface of the earth ($\mathbf{I_{R_{\alpha-}}}$), and that portion of the sun's energy that first reflects off the surface of the earth and then gets Rayleigh scattered ($\mathbf{I_{R_0}}$).

Finally, the satellite sensor may receive energy directly from the earth's surface. The energy level of the radiation leaving the surface (whether the earth's natural temperature-related radiations or reflection of solar or man-induced signals) is denoted by $\mathbf{I_0}$.

Transmissivity. Since the energy received by a remote sensor must propagate through the atmosphere, this energy may be changed during transit due to the different scattering modes and the absorption and reradiation of energy by the components of the atmosphere, all of which are highly frequency-dependent. The effect of the atmosphere on electromagnetic propagation is modeled by a *transmissivity* factor denoted by $\tau_{(\lambda)}$. Figure 6-4 shows the transmissivity of radiation through the atmosphere as a function of percent transmission versus wavelength.

The figure also indicates the atmospheric constituents responsible for the poor transmittance of some frequencies through the absorption mechanism described earlier in Chapter 4 using a similar curve showing the spectral energy propagation through the atmosphere. Figure 6-4 only

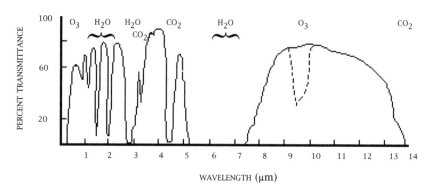

Figure 6-4. Atmospheric transmissivity. Some atmospheric constituents react with certain wavelengths, absorbing some or all of the associated energy.

shows a portion of the infrared spectral band and not the entire electromagnetic spectrum. Frequency ranges with poor transmissivity are known as *absorption bands,* while regions where radiations pass through relatively unaffected are called *windows.* As an example, we can see well during the day because the atmosphere is practically transparent to the visible frequency range; but a couple of hours at the beach doesn't burn us too badly because most of the ultraviolet radiations of the sun are absorbed high in the atmosphere. A particular remote sensor's purpose may be to observe and measure some phenomenon at the earth's surface, in which case it must negotiate frequencies with significant atmospheric absorption and use the window frequencies. Many sensors, sometimes called *sounders,* wish to measure atmospheric or ionospheric conditions and will intentionally select frequencies of observation that interact with and measure high-altitude phenomena.

The component of the total energy received by the remote sensor from just the *surface* is given by:

$$I_{o(\theta)} = I_o e^{(-\tau_{(k)}/\cos(\theta))}$$

where

$$I_o = F_{(k)} e^{(-\tau_{(k)}/\cos(\theta_o))}$$

which shows the dependency of the received energy on the spectral output of the sun $F_{(k)}$ and the effects of the transmissivity $\tau_{(k)}$ on the energy propagation. Similar equations exist for the other components of the total received energy which are also dependent on these factors. The total energy received by the remote sensor is the sum of these radiations which, when expressed in equation form, constitute the *radiative transfer equation:*

$$I_T = I_{o(\theta)} + I_{R_{(tot)}} + I_A \qquad (6\text{-}7)$$

where the total Rayleigh scattered input is given by $I_{R_{(tot)}}$, and I_A is the received aerosol scattered component.

The reduction of this raw received information is quite complicated and requires some prior knowledge of the behavior of the atmosphere. Over many years, algorithms have been developed which are used to reduce the unrefined sensor source or raw data and identify the physical processes which contribute to the total energy received by the remote sensor. Algorithm development is based on a theoretical understanding of the physical processes involved, bolstered, and refined by empirical experiments using aircraft and satellite remote sensor observations. From this, the basic geophysical properties discussed earlier, such as temperatures, atmospheric constituents, and land resources, can be distilled.

MEASUREMENT DIMENSIONS OF REMOTE SENSORS

All remote sensors observe in the context of specific spectral regions, and within each spectral band, a sensor further defines an observation in terms of space, time, and signal strength. The dimensional extent of a sensor's ability to observe has been defined as the swath width or field of view (FOV). Within the FOV, a sensor's spatial acuity is defined within the IFOV elements (pixels). Each IFOV element is further defined in terms of the strength of the signal identified with a specific observation or scene location. Considering all of the IFOV elements that combine to constitute a scene, the variation in signal strength produces contrast. Sensors depend on a satellite's orbit and the sensor's scan pattern to determine the precise observation time and sequence for a particular earth location.

Spatial Performance

Passive Sensor Spatial Resolution. The spatial performance of passive sensors is defined by the wavelength scale dimension of the focusing aperture (in radians) and the range between the observing sensor and the observation area. The satellite orbit and sensor scan pattern can alter the exact dimensions of the earth-projected sensor IFOV, but the beam projection is fundamentally diffraction limited. Active sensors can go a step further in gained improved resolution in both the range and along-track dimensions.

Active Sensor Spatial Resolution. Active sensors can improve on the resolution in the *range* or beam direction by controlling the length of the pulse of energy transmitted by the sensor. This is known as *pulse ranging* and is shown in Figure 6-5.

The active sensor sends out a pulse of energy with a *pulse length* of τ seconds. This pulse propagates in the direction of the beam (boresight) and illuminates the IFOV area, after which some of the energy reflects back to the receiver. In this case, the range resolution (R_r) depends on the pulse length and the signal propagation velocity (c_m) which is approxi-

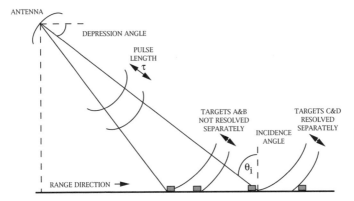

Figure 6-5. Active sensor range resolution. Active sensors can improve on resolution in the range direction by controlling the pulse length of the transmitted energy.

mately the speed of light for most active sensor frequencies and space-to-earth viewing scenarios. The relationship is given in equation 6-8.

$$R_{r_{slant}} = \frac{\tau}{2} c_m \qquad (6\text{-}8)$$

This sensor resolution is achieved in the slant range plane and must be translated into the ground plane as a function of the local *target incident angle* (θ_i) as shown in equation 6-9.

$$R_{r_{ground}} = \frac{\tau \, c_m}{2 \sin \theta_i} \qquad (6\text{-}9)$$

Radars and lasers improve their spatial resolution in the range or beam direction by using this technique. Active radar sensors can either stop at this point or continue this spatial resolution improvement process by devising a method of improving the along-track or azimuth spatial performance of the sensor. The two basic types of active imaging radar systems are *real-aperture radar* (RAR) and *synthetic-aperture radar* (SAR), which differ primarily in the method of achieving a desired resolution in the azimuth direction. The RAR obtains its along-track resolution through the wavelength scale physical dimensions of its antenna in the along-track direction. As a result, the RAR uses an antenna of the maximum practical length to produce a narrow, diffraction-limited, angular beam width in the azimuth direction.

Synthetic-aperture radar systems employ a smaller antenna that transmits a relatively broad azimuth or along-track beam as shown in Figure 6-6. A SAR "synthetically" obtains its along-track resolution by moving the real antenna beam in relation to the target. In the most basic sense, there must be a translation of either the target through the real beam, the real beam through the target, or a combination of both processes. As shown in the figure, as the radar beam translates the target, the radar sends a string of carefully prescribed pulses illuminating and receiving backscatter responses from the target area. The SAR preserves and saves the amplitude and phase histories of each of these responses. A special image signal processing system performs a target specific weighting, shifting, and summing process focusing on one resolution element at a time and, as a result, achieves an along-track resolution that is independent of range and wavelength, becoming equal to one-half of the actual antenna length in the along-track direction, as shown in the figure.

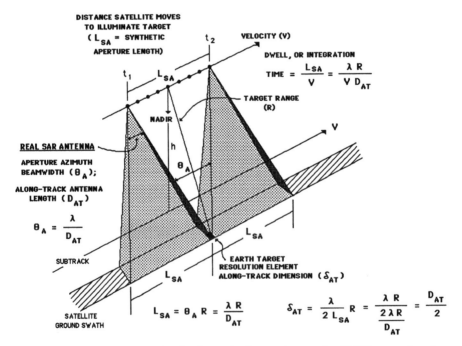

Figure 6-6. Aperture synthesis. A synthetic aperture radar (SAR) uses signal processing over many radar returns to obtain resolutions comparable with normal radars of much larger apertures.

Spectral (Radiometric) Performance

The *spectral* performance of a sensor depends on the frequency (or wavelength) of the radiation which the sensor is designed to detect. The wavelength(s) used to make a measurement is selected based on the radiative (*radiometric*) properties of the *target* and/or on other considerations, such as interaction with or avoidance of the atmosphere or ionosphere. For instance, in clouds, water occurs as aerosol-sized particles of liquid rather than vapor. Clouds absorb and scatter electromagnetic radiation at wavelengths less than about 0.3 cm. Sensors are designed to take advantage of this effect to measure atmospheric characteristics. Only radiation of microwave and longer wavelengths is capable of penetrating clouds without being significantly scattered, reflected, or absorbed. These (radar) frequencies are commonly used when information below cloud layers is desired.

The spectral performance of a system is a measure of how well the system uses these frequencies (or combinations of frequency ranges) to extract information on a desired target. Much of the success of a remote sensor depends on knowledge of the radiometric properties of the target. These target/scene characteristics determine the relative strength or intensity of observed signals based on their reflectance, absorption, or other properties for the particular frequencies involved.

Temporal Performance

The *temporal* performance of a remote sensor is a function of the repeated observations of a phenomenon based on the expected rate of changes in that phenomenon. Global winds change on an hourly or daily basis while changes in the earth's geodetic shape or magnetic field may take years or millennia to occur. The observation rate of a specific phenomenon or area is a function of the orbit in which the satellite is placed and the sensor FOV produced by the scan pattern. An orbit and scan pattern can be preferentially designed to create a revisit sequence and local time observation tailored to a particular application.

REMOTE SENSING SATELLITES

Early remote sensing satellites were technology and lift capacity constrained, resulting in relatively simple sensors designed to observe singular regions of the electromagnetic spectrum. Modern satellites tend to maximize in-orbit returns by investigating several electromagnetic regions simultaneously. Figure 6-7 shows the frequency bands used by modern remote sensing systems, with the regions of interest aligned with particular frequency ranges. Note that radiation in the short wavelength portion of the IR band and the long wavelength portion of the UV band is detectable by photographic film. Therefore, these wavelengths are included with the visible wavelengths and are designated as the photographic remote sensing band.

In conclusion, it should be noted that mission application objectives and remote sensors drive the spacecraft and orbit design. The sensor systems are designed first to give the desired performances as described above, and the other spacecraft systems (such as attitude control, power, and communications, which will be discussed in a separate chapter) are created to support these sensors in the collection and transmission of the observed data to the ground for synthesis, geophysical analysis, and dissemination.

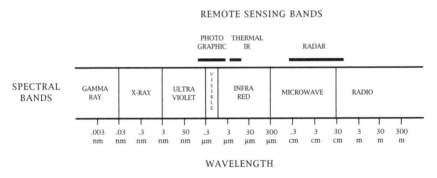

Figure 6-7. Remote sensing bands. Modern remote sensing satellites may contain multiple sensors designed to investigate several areas within remote sensing bands simultaneously.

REFERENCES/ADDITIONAL READING

Cantafio, L., *Space-Based Radar Handbook*. Boston: Artech House, 1989.

Sabins, F., *Remote Sensing Principles and Interpretation*. 2nd ed. New York: W.H. Freeman and Company, 1985.

Schnapf, A., *Monitoring Earth's Ocean, Land, and Atmosphere from Space—Sensors, Systems and Applications*. Washington, D.C.: American Institute of Aeronautics and Astronautics, 1985.

Manual of Remote Sensing, Volumes I and II, 2nd ed. Washington, D.C.: American Society of Photogrammetry, 1985.

Beal, R., *Spaceborne Synthetic Aperture Radar for Oceanography.* Baltimore: The Johns Hopkins University Press, 1981.

EXERCISES

1. A sensor is to be placed into geostationary orbit to monitor its field of view constantly for infrared signals, indicating possible ICBM launches. The sensor is to consist of a 10 × 10 pixel square array of photodetectors looking at the frequencies associated with temperatures between 1,500 and 3,000° K (1.29 μm median wavelength). The electronic output of the photodetectors is a sinusoid with a changing amplitude that represents the detected temperature. This signal is to be digitized using 16 bits per word with a sampling frequency of 1 KHz.

   Draw a sketch and determine the geometry of the situation, including look angle (θ_n), earth angle (γ), and elevation angle (ψ), for the maximum slant range (R_{max}). (Hint: $\psi = 0$ at R_{max}.) Determine the size of each photodetector required to give a ground resolution (the IFOV for one detector) of 10 km when looking straight down. What is the difference between this resolution and that when the sensor is looking out at R_{max}? Compute the system IFOV and GIFOV at nadir and at R_{max}. Also, compute the temperature sensitivity and transmission rate of the system. Finally, discuss how you would attempt to scan the entire visible earth's surface for complete coverage. Based on your scan method and the given sampling rate (1 sample = 1 area covered), estimate how long it would take your system to scan the entire visible surface once (assuming no overlap between areas scanned each sample).

2. Determine the size aperture a remote sensor would require if it were looking for possible ICBM temperatures between 1,500 and 3,000 °K (1.29 μm median wavelength) and wished to see the entire earth's surface possible at all times from its geostationary position while always looking straight down at nadir. Use Figure 6-1 to work out the geometry of the problem. Hint: To do this you need to find the distance perpendicular from the nadir line (between the satellite and the center of the earth) and the point on the horizon of the satellite's field of view.

3. If the sensor in Exercise 2 is a 10×10 pixel array of CCDs generating byte-sized words each sample, and the transmission rate is limited to 200 Mbps, determine the sensitivity of the system and the maximum sampling frequency for the system.

4. A photographic reconnaissance satellite operates at an altitude of 150 km, always looking straight down. It has a lens with a diameter of 0.5 meters which projects the received images onto a 64×64 pixel array of CCDs with a receiving capability from 5×10^{14} Hz to 6×10^{14} Hz (0.6å to 0.5å in wavelength where 1å = 1 micron = 10^{-6} meter). Pulse code modulation (PCM) is used to digitize the data, but the system is limited to 16 bits of information for each CCD each sample. Fifteen thousand pictures are taken each second (for overlap). Determine the following:

   a. The field of view of the image. Use the median wavelength of 0.55å for the calculation.

   b. The number of quantization levels available.

   c. The frequency sensitivity of the system.

   d. Total number of bits produced each sample.

   e. Transmission rate in Mbps.

   f. (EXTRA CREDIT) If the CCD array is 1 square meter, what is the image resolution of the system? (OR) Determine if the picture-taking frequency is sufficient for image overlap.

5. A remote sensor has a 32×32 pixel array of infrared-sensitive charged-couple devices (CCDs). The range of sensitivity for the CCDs is centered between 250 °K and 500 °K. Pulse code modulation (PCM) is used to digitize the data, and the sensitivity range has been divided into 16 quantization levels. Sampling frequency is 10 Khz.

   a. What is the temperature sensitivity (represented by each Q-level)?

   b. How many bits of information does a CCD produce each sample?

   c. What is the total number of bits produced each sample by the array?

   d. What is the transmission rate of the system in bits per second (bps)?

6. A passive remote sensor has a 1 meter diameter circular lens and operates at a frequency of 3×10^{14} Hz. Calculate the IFOV at a range of 1,000 km. What would be the ground IFOV if the grazing angle was 60°?

7. A remote sensing satellite is operating at an altitude of 800 km. The maximum nadir look angle of the scanning sensor is 30° side to side relative to the satellite's ground track. Calculate (a) the grazing angle and (b) the slant range corresponding to the maximum look angle, and (c) the total horizon-to-horizon FOV distance in km.

8. An active radar sensor sends a pulse of 3×10^{-9} sec (3 nanosec) which strikes the earth at a 30° incidence angle. Calculate the ground resolution in meters of the radar.

CHAPTER 7

Satellite Navigation

The concept of using satellites for navigation dates back to *Sputnik 1* in 1958 when engineers monitoring this first orbiting body observed that the substantial Doppler frequency shift of the received telemetry signal could be used to determine accurately the satellite's position and orbit. It was quickly shown that the inverse situation was also possible; that if the satellite position (orbit) were known accurately, an observer's location on the earth could be determined. After several years of work at the Applied Physics Laboratory (APL) of Johns Hopkins University, a prototype satellite was developed and launched, and satellite navigation was born.

The first operational systems were part of the Navy Navigation Satellite System known as "Transit." The system uses the Doppler positioning method described further in this chapter and requires information gathered from only a single satellite, over the time of passage overhead, to produce a fix. The newest generation of navigation satellites are part of the Global Positioning Satellite (GPS) system. This system uses a pulse ranging method of positioning and requires four satellites to be in view of the receiver simultaneously for a complete solution. Though the requirements are more stringent, GPS provides real-time, three-dimensional position, track, and speed with accuracies many times better than the earlier Transit system.

POSITION DETERMINATION USING DOPPLER TECHNIQUES

The original method of determining position from satellite signals is through observation of the Doppler shift in frequency of a known signal as the satellite passes over the observation area. The shift in frequency is a unique function of the motion of the satellite relative to the observer's position. To compute an observer's location accurately, the position and orbit of the satellite and the frequency of the transmitted signal must be known. Also, timing between the transmitting satellite and the observer's receiver is important. Most of these requirements are provided by a message contained in a separate signal transmitted by the satellite (the *navigation message*), which reports the satellite *ephemeris* (position and orbit information) and a timing signal. The frequency of transmission is known by all users and must be provided by a very stable source aboard the satellite. The following sections show how this information is combined to allow an observer to compute a position on the earth.

Doppler Ranging

The frequency received from the satellite (f_R) consists of the transmitted frequency (f_T) plus a Doppler frequency ($\pm f_D$) due to the relative motion between the satellite and receiver. The receiver also has a stable oscillator which produces a reference signal at approximately the same frequency as transmitted by the satellite (f_G). Subtracting this frequency from the received frequency, the receiver determines the shift in frequency in terms of a *Doppler count*. The Doppler count represents a counting of the number of frequency cycles occurring between timing signals transmitted by the satellite. In equation form:

$$N_1 = \int_{t_1 + R_1/c}^{t_2 + R_2/c} (f_G - f_R)\, dt \tag{7-1}$$

where $t_1 + R_1/c$ represents the time the signal was received after it was transmitted from the satellite at time t_1 and traveled to the receiver over a slant range of R_1 at the speed of light c. Expanding equation 7-1 into two parts:

$$N_1 = \int_{t_1 + R_1/c}^{t_2 + R_2/c} f_G\, dt - \int_{t_1 + R_1/c}^{t_2 + R_2/c} f_R\, dt \tag{7-2}$$

The first part is simply the integral of a constant since the receiver produces a stable frequency (f_G), but the second part contains the changing received frequency f_R. However, the second integral also represents the number of cycles *received* between the two timing signals (sent by the satellite) which must be equal to the number of cycles *sent* by the satellite between those same two times. This allows us to rewrite the second part in terms of simply the transmitted frequency and the transmitter timing signals:

$$N_1 = \int_{t_1 + R_1/c}^{t_2 + R_2/c} f_G \, dt - \int_{t_1}^{t_2} f_T \, dt \qquad (7\text{-}3)$$

Because f_T is also considered constant and known, both the above integrals can be performed resulting in:

$$N_1 = f_G \left[\left(t_2 + \frac{R_2}{C} \right) - \left(t_1 + \frac{R_1}{C} \right) \right] - f_T \, (t_2 - t_1) \qquad (7\text{-}4)$$

By combining terms, equation 7-4 can be rewritten as:

$$N_1 = (f_G - f_T)(t_2 - t_1) + \frac{f_G}{C}(R_2 - R_1) \qquad (7\text{-}5)$$

The first term of equation 7-5 is just the difference between the receiver's oscillator frequency and the satellite's transmitted frequency, which is a constant term and can simply be ignored. The second term is a measure of the change in range between the satellite and the receiver during the time between two timing marks. This information is used to determine the receiver's position. Notice that this range difference is given in terms of wavelengths of the *receiver's* generated frequency (f_G).

Computation of Position

Computation of position usually requires a small digital computer. The computer determines the satellite position at each timing signal from the ephemeris information transmitted by the satellite. The computer also uses an approximate position of the receiver, either inputted by the user or estimated from the last fix. The expected change in slant range between the satellite and the receiver (based on the estimated position) is computed and compared to that actually measured. The estimated position is then shifted in such a way that the differences are decreased. Through several iterations of this procedure, the estimated position is refined until the differences are reduced to an acceptable value.

As mentioned earlier, the Transit system requires only a single satellite to be in view to determine position. The iterative process is repeated for as long as the satellite is in view resulting in an increasing positioning accuracy with time, as illustrated in Figure 7-1. If the receiver is in motion, the computer must know the motion accurately to compute the Doppler-based

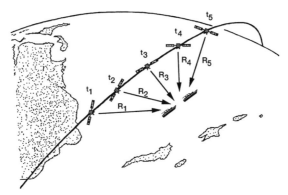

Figure 7-1. Doppler positioning. Successive position iterations may be achieved from a single pass of a single Transit satellite.

slant range change precisely and to know the change in receiver position over the observation time. Multiple fixes can be combined to provide position, track, and speed for subsequent iterations resulting in increased accuracy. If a single satellite is used, a dead reckoning (DR) plot is used between times the satellite is below the horizon to give an estimated position, track, and speed for the next satellite encounter.

Errors and Accuracies

The theoretical accuracy of the Doppler ranging method is associated with the wavelength of the frequency used for the navigation signal. For instance, a 400 MHz signal (currently used by Transit) corresponds to a 0.75 meter wavelength and a similar *theoretical* accuracy (for a single fix). Unfortunately, many sources of errors exist which affect the accuracy of satellite navigation systems.

Refraction Errors. There are two sources of refraction errors as the satellite signal propagates between the satellite and the receiver. The first is introduced by the ionosphere which changes the path of propagation of the signal as described in Chapter 5. This change introduces a shift in the frequency of the signal which would affect the determination of the slant range change between the satellite and receiver. In order to compensate for this shift, navigation satellites transmit over two different frequencies and compare the Doppler measurements made at each to reduce the error introduced by the ionosphere.

The second source of refraction error is the troposphere which affects the speed of propagation of the signal as it travels through the atmosphere. This error changes with atmospheric changes, such as temperature and humidity, and with elevation angle between the receiver and the satellite (signals travel through more of the atmosphere, and are affected more, with lower elevation angles). The troposphere affects all frequencies similarly and is not as easily subtracted as the ionospheric effects. Receiver station compensation for tropospheric errors, if any, is usually done using simple atmospheric models.

Position Errors. The accuracy of the fix obtained from Doppler positioning, with respect to ground latitude and longitude, is a function of the receiver's position with respect to the satellite orbital plane and knowledge of the receiver's altitude above the reference "spheroid" (the surface defined as if the earth were actually a perfect sphere) on which satellite position is based. (Note: The "geoid" is another common reference surface and represents the surface defined by mean sea level worldwide. Both the spheroid and the geoid are different from the actual topography of the earth. Knowing one's altitude, in this sense, is not as simple as it may seem.) Figure 7-2 shows how the same range rate information would report a different position if the altitude were not accurately known. Notice that, for the same error in altitude, the error in computed position is greater if the receiver is closer to the satellite orbital plane.

Additionally, as was mentioned earlier, receiver position, track, and speed are required for accurate positioning. Estimated position is not as important as track and speed, as this error will be eliminated through the iteration procedure of the computer's correlation with slant range rates. However, track and speed are very important as they affect the computation of the Doppler shift between satellite and receiver.

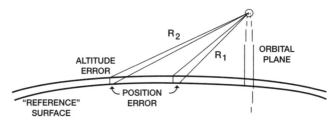

Figure 7-2. Altitude error. Computed position may have an error associated with uncertainty of "altitude" with respect to the reference plane.

Satellite Errors. Errors attributable to the satellite include rounding in the ephemeris data transmitted by the satellite to the receiver, and inaccuracies of this reported information due to orbital parameters such as atmospheric or solar pressure drag and inaccuracies of prediction of the satellite orbit over time due to insufficient knowledge or modeling of the geopotential (gravitational) model of the earth. Satellite ephemeris is computed frequently by ground stations and the navigation message updated often to reduce these errors. Additionally, satellite and receiver clock and oscillator instabilities may introduce another, not insignificant error.

PULSE RANGING AND PHASE DIFFERENCE POSITIONING

Experience with the first Doppler-based satellite navigation systems produced newer ideas for conducting positioning from orbiting satellites, which offered increased accuracies and eliminated some of the sources of errors affecting the earlier method. Doppler methods compare estimated range rates with those derived from the Doppler shift to determine a position correction. A simpler concept involves deriving one's actual range directly from the satellite transmitted signal. The idea is depicted in Figure 7-3.

If a signal (a pulse, for instance, as shown in the figure) were sent out by a satellite at a known time, the exact range from the satellite to the receiver could be found from:

$$R = c_m \Delta t \qquad (7\text{-}6)$$

where c_m is the speed of propagation of the signal and Δt is the time between transmission (t_{xmit}) and reception (t_{recv}) of the signal. The range represents a sphere, somewhere on which the observer would lie. An exact position, in three dimensions, could be obtained by the intersection of

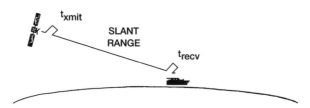

Figure 7-3. Pulse ranging. Distance (slant range) between a transmitter and receiver can be determined by knowing the time it takes a signal to travel.

three of these range spheres (once again, knowing precisely the position of the transmitter[s]).

The concept is quite simple, but actual implementation of the idea poses some real problems. For example, the signal sent by the satellite, from which the observer will compute range, must allow the receiver to determine exactly *when* a *particular event* was transmitted.

Ranging Signal

Specifying the particular "event" to be used to compute range lends itself very well to digital techniques. To determine exactly which "event" has been received (assuming the satellite transmits continuously), a digital code with a unique pattern could be used. The receiver picks up the signal and compares it to the same code stored in memory to find where along the pattern the received information is located.

A convenient method for impressing the digital code onto a suitable carrier wave is through biphase modulation (review Chapter 5 if necessary). The "event" is a change (or no change) of the phase of the carrier wave by 180° each time the *state* of the digital code changes. An example of a biphase modulated signal is shown in Figure 7-4.

The transmission rate of the digital sequence determines two important characteristics of the system. A small time between phase changes (t_b) allows the receiver to more accurately determine range from the transmitter, but a larger pulse period ensures better detection of the pulse and error-free use of the information. Tradeoffs between these objectives results in the modulation frequency used (which affects the receiver minimum signal-to-noise ratio) and also determines the theoretical accuracy of the system.

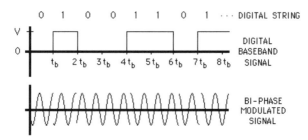

Figure 7-4. Biphase modulated signal. An "unrepeating" code is modulated onto a carrier frequency to determine the timing "mark" for pulse ranging.

Timing Signal

Another difficulty posed by the ranging concept is determining exactly *when* the particular "event" was sent. Solving this problem involves two steps: *correlation* and *synchronization*.

Correlation. Figure 7-5 displays the three signals involved in pulse ranging. The first signal is the biphase modulated signal transmitted by the satellite ($S_{o(t)}$). Below it is the identical signal stored in memory of the receiver ($S_{1(t)}$). Note that this stored signal may not be exactly synchronized in time with the transmitted signal. The third signal shown ($S_{2(t)}$) is the received signal (ignoring such distractions as noise).

Both the replica signal and the received signal are digitally sampled and then the signals are compared using a digital *cross correlation* function of the form of equation 7-7:

$$R_{(t)} = \sum_t [S_{1(t)} \times S_{2(t+t)}] \qquad (7\text{-}7)$$

(Note: Summation in discrete-time [digital] mathematics is the same as integration in conventional math.) Summation over many values of **t** results in a particular time (t_0) for which the summation is maximum. This represents the *offset time*, or time between the occurrence of the same particular event (bit/phase change) in the received signal and the stored replica code.

Synchronization. The offset time found above does not yet represent the true Δt between transmission and reception of the signal, as the satellite

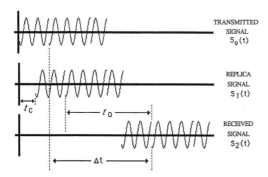

Figure 7-5. Pulse ranging signals. The relationship between these signals is used to determine the times required for an accurate ranging.

and receiver clocks may not be in perfect synchronization. This *clock off-set time* is shown in Figure 7-5 sketch as the t_c time difference between the satellite signal and the receiver's replica code. If it is assumed that the satellite is the absolute time reference, then this clock offset time would produce an error in range if not allowed for. The method for determining the value of the clock offset time is described as part of the ranging process described next.

Range Determination

The range equation (eq. 7-6) would now look like:

$$R = c_m \, (t_o \pm t_c) \tag{7-8}$$

In terms of an X, Y, Z coordinate system, this equation would be of the form:

$$R_1 = [(X_{1(t_1)} - X)^2 + (Y_{1(t_1)} - Y)^2 + (Z_{1(t_1)} - Z)^2]^{1/2} \pm c_m(t_c)$$

where the $X_{1(t_1)}$, $Y_{1(t_1)}$, $Z_{1(t_1)}$ terms represent the position of the transmitter (satellite) at the time of transmission (t_1). **X, Y,** and **Z** represent the position of the receiver computed from the range information derived from the correlation offset time $c_m \cdot t_o$, and $c_m \cdot t_c$ represents the range error due to the satellite/receiver clock offset time.

Assuming that the position of the satellite is known precisely at all times (via a navigation message as described earlier), this equation has four unknowns involving the **X, Y, Z** position coordinates of the receiver and the receiver clock offset time t_c. To solve for these unknowns, at least four range signals must be available at the same time to solve four similar equations simultaneously:

$$R_1 = [(X_{1(t_1)} - X)^2 + (Y_{1(t_1)} - Y)^2 + (Z_{1(t_1)} - Z)^2]^{1/2} \pm c_m(t_c)$$

$$R_2 = [(X_{2(t_2)} - X)^2 + (Y_{2(t_2)} - Y)^2 + (Z_{2(t_2)} - Z)^2]^{1/2} \pm c_m(t_c)$$

$$R_3 = [(X_{3(t_3)} - X)^2 + (Y_{3(t_3)} - Y)^2 + (Z_{3(t_3)} - Z)^2]^{1/2} \pm c_m(t_c)$$

$$R_4 = [(X_{4(t_4)} - X)^2 + (Y_{4(t_4)} - Y)^2 + (Z_{4(t_4)} - Z)^2]^{1/2} \pm c_m(t_c)$$

Note that the same receiver clock offset time t_c appears for each of the received signals. This assumes that each of the received signals comes from satellite transmitters synchronized exactly with absolute time. If this were the case, then even if the above ranges were computed from four different satellites, when the information was received all the times would be the same, i.e., $t_1 = t_2 = t_3 = t_4$. In practice, another time correction (included in the navigation message) is incorporated to synchronize the transmitters on board the different satellites themselves to an absolute reference time.

Errors and Accuracies

The refraction-type errors described earlier affect all types of transmissions through the atmosphere and decrease the accuracy of the ranging process as well. As with Transit, GPS satellites transmit on two different frequencies to compensate for ionospheric effects on the signals. These much higher frequencies (1.58 and 1.23 GHz) are also less affected by the troposphere and additionally allow a much higher bit rate resulting in greater accuracies.

No estimated position or track and speed information is required because solution of the four range equations simultaneously gives a very accurate three-dimensional fix. Satellite errors can be minimized with more accurate and stable oscillators, better geopotential models and more stable orbits, and updating of the navigation message more frequently.

Table 7-1 compares the accuracy of some of the more common navigation systems. TACAN (Tactical Air Navigation), Omega, and Loran are examples of existing ground-based radio navigation aids.

Table 7-1
Accuracy of Some Common Navigational Systems

System	Position Accuracy (meters)	Velocity Accuracy (m/sec)	Comments
TACAN	400	none	Line of sight air navigation.
Omega	2,200	none	Worldwide radio navigation.
Loran-C	180	none	Localized area (U.S. shore) radio navigation.
Transit	200	none	Worldwide, but up to 100 min. between satellite passes.
GPS	15*	0.1*	Global, 24-hr, all-weather availability.

*Accuracies only available to authorized (military) users. Public accuracies are approximately a magnitude less.

REFERENCES/ADDITIONAL READING

Logsdon, T., *The Navstar Global Positioning System.* New York: Van Nostrand Reinhold, 1992.

Dutton, L., *Military Space.* Great Britain: Brassey's, 1990.

McElroy, J., *Space Science and Applications.* New York: IEEE Press, 1986.

Cochran, C., *Space Handbook.* Alabama: Air University Press, 1985.

EXERCISES

1. List and expound upon some of the benefits achieved by using navigation satellites.
2. Assuming a slant range to a GPS satellite of 20,000 km and a speed of light of 3×10^8 m/sec, determine the range inaccuracy associated with a 1% error in time measurement between transmit and receive of a timing signal.
3. Using the parameters above, determine the time measurement accuracy (percentage and number of seconds) required to achieve a 50-meter slant range measurement capability.

PART 3

Spacecraft Systems and Design

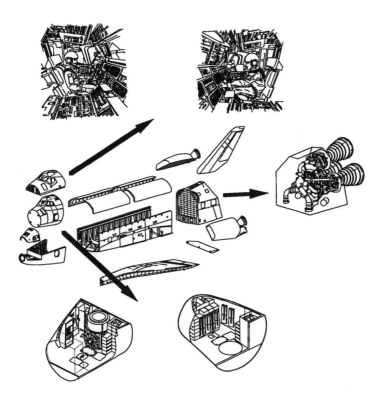

CHAPTER 8

Spacecraft Systems

The preceding chapters have discussed many of the space science topics involved with the operation of spacecraft and spacecraft applications. The purpose of this chapter is to describe, in more detail, the individual elements which make up a complete space system. Before continuing, we must first define precisely what is meant by a space system. The spacecraft itself is part of a larger system, which includes ground-based telecommunication, tracking, data collection, processing, and dissemination elements that are described in the chapter under the heading of Ground Support Systems. Other space-based assets such as communication relay satellites used to link spacecraft data to the ground systems and navigation satellites that provide precise geopositioning information to orbiting satellites can also be an essential part of the larger architecture of our spacecraft system. This constellation of interlocking space-based and ground-based systems are all tied to a system user with a mission goal. Whether the mission applications goal is scientific, military, civil, or commercial, the over-arching space system, with networked space and ground elements, begins and ends with the user. The user establishes the applications goals and operating protocols at the beginning of the system development process and makes use of the communications, remote sensing, and navigation information produced by the system when it is in place. Our description of this complex end-to-end system begins with a discussion of the spacecraft element by outlining the rudiments of satellite design.

SATELLITE DESIGN

Figure 8-1 identifies the basic elements of a satellite system. In addition to the payload, satellite, and launch vehicle, there are many devices that must be added to the system to complement the payload, such as special downlinks, tape recorders, and cryogenic coolers. The mission and payload characteristics also place requirements upon the satellite bus subsystems to provide special capabilities in each of the subsystem areas, for

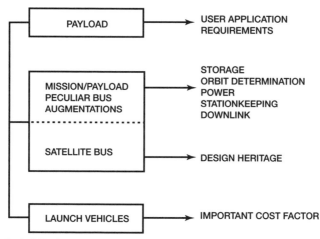

Figure 8-1. Mission requirements. Many factors affect spacecraft design.

example, precise pointing and stationkeeping, and autonomous operation for extended periods of time. This theme is further expanded in the following sections that follow the design process illustrated in Figure 8-2 which identifies the salient properties of the mission and payload that drive the design of satellite bus subsystems:

1. Attitude Reference and Control (ARC)
2. Power
3. Thermal Control
4. Orbital Maintenance (station-keeping)
5. Propulsion
6. Data Handling
7. Onboard Computer
8. Telemetry Tracking and Control (TT&C)
9. Structure.

In the design process, these subsystems are interdependent, and any design must be iterated so as to take into account interlocking design architecture, power, envelope, and mass interactions.

The origins of satellite design are mission and payload characteristics. Mission characteristics establish the environment that the satellite will be subjected to during its lifetime in space as well as the earth observation revisit frequency, range from the satellite to earth, velocity, and other important characteristics, all of which are directly related to the payload

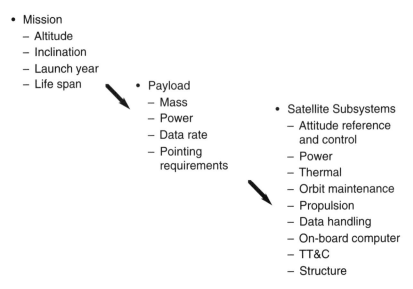

- Mission
  - Altitude
  - Inclination
  - Launch year
  - Life span

- Payload
  - Mass
  - Power
  - Data rate
  - Pointing requirements

- Satellite Subsystems
  - Attitude reference and control
  - Power
  - Thermal
  - Orbit maintenance
  - Propulsion
  - Data handling
  - On-board computer
  - TT&C
  - Structure

Figure 8-2. Satellite design flow diagram.

application. Payload attributes such as dynamic mass and envelope dimensions, data collection and transfer rates, power profiles, and pointing and stabilization requirements also determine the design requirements for the satellite bus subsystems.

Mission

Specification of the six orbital elements is determined by user applications requirements concerning temporal (scene revisit frequency), field of view, solar illumination geometry, and other important considerations. Part of the mission design process includes decisions about the launch vehicle's role in achieving the final orbit and how much propulsion the satellite design must provide for the satellite to reach its final destination.

Specifically, the mission affects the entire design because it establishes the environment and duration for the satellite's lifetime. The orbit influences the power subsystem by setting the eclipse times and mission lifetime that the solar panel output must comply with; the orbital maintenance subsystem by determining the neutral density environment that causes drag and the time duration that the satellite will be exposed to this orbit-modifying force; and the satellite propulsion subsystem by determining the Δv necessary to achieve final orbit.

Payload

The payload mass and mass dynamics (e.g., scanning antennas and spinning mirrors), pointing accuracy, dynamic envelope dimensions, data and power operating profiles, and other characteristics propagate into many complex aspects of the satellite design.

For example, the mass of the payload sets the pace for the calculation of the on-orbit mass of the entire satellite, thus setting the goal for the propulsion subsystem Δv. The ARC subsystem keys on the pointing accuracy required by the most demanding payload element and uses this value to determine the candidate control architectures capable of achieving this accuracy, then goes on to compute the control system's mass required to stabilize the satellite in its orbit. The on-off duty cycle and power demands of the payload create the profile for the power subsystem that must collect and store energy as the satellite goes in and out of (eclipses) a view of the sun. The thermal subsystem considers eclipse and sun periods and satellite orientation along with the mass and power requirements of the payload to arrive at an estimate of the thermal subsystem mass. The orbital maintenance subsystem uses the mass of the payload to drive the total on-orbit mass and converts this into a satellite design envelope yielding a ballistic cross section. This estimation is necessary to compute drag (which varies with the environment predicted during the expected mission lifetime), which in turn determines the fuel necessary to maintain the orbit over the expected mission lifetime. The TT&C and data handling subsystems key on the payload data rate and duty cycle to determine the need for downlink bandwidth and the type and capacity of onboard data recorders. The structure subsystem evaluates the needs of the satellite using both static and dynamic mass estimates from the payload and other subsystems to compute the total structural mass.

Payload requirements influence and can also be limited by satellite performance attributes such as pointing accuracy, structural rigidity, power, and thermal requirements. These and many other important considerations are cross-linked as the designer proceeds from mission and payload specification to a review of each of the satellite bus subsystems.

Launch Vehicle and Site Selection

The launch vehicle may be selected prior to satellite bus subsystem design or after the design has been completed. If the former option is chosen, the design proceeds down a path of launch vehicle optimization, and a constant tally of information in the form of mass, envelope, and power "vital statistics" must be tracked indicating the degree of launch vehicle capacity, e.g., envelope and lift, used as the design proceeds through the satellite bus subsystem analysis. Conversely, if the satellite bus subsystem design proceeds without a launch vehicle first being selected (the case for a preselected "common bus" heritage design path), then the resulting satellite design must be matched to an available launch vehicle. Figure 8-3 identifies a few candidate launch vehicles that are available to spacecraft designers. Table 8-1 provides an example of a typical launch performance database for the Pegasus XL that can be launched from an aircraft platform. Pegasus is a leading launch vehicle candidate for smaller satellites such as SMALLSATs and LIGHTSATs.

The following summary discussion of each of the satellite bus subsystems serves to introduce the more detailed elements of design associated

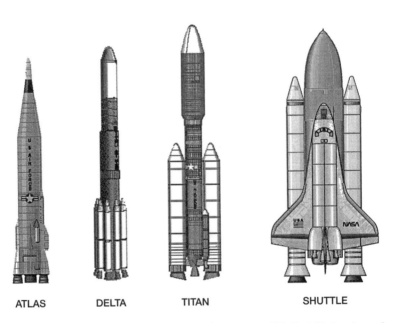

ATLAS   DELTA   TITAN   SHUTTLE

Figure 8-3. Launch vehicles. While the major types of United States launch vehicles are shown, there are several smaller U.S. launchers available as well as several foreign vehicles.

Table 8-1
Typical Launch Performance Database for
Pegasus XL Aircraft

Lift Capacity (kg)	Inclination (degrees)			
Altitude (km)	0	28.5	90	98
200	410	395	295	280
400	390	375	275	260
600	370	355	255	245
800	345	335	240	230
1,000	325	310	225	210
1,200	305	290	205	190
1,400	280	265	190	170
1,600	260	245	175	155
1,800	235	225	155	140
2,000	210	205	135	115

Fairing (L × Dia, m): 1.7, 1.1
Launch Cost ($M): 13
Pegasus data from "Commercial Pegasus Launch System Payload User's Guide" (Rel. 3.00, 1 Oct 1993).

with each subsystem. The designer has many choices of design architecture, including types of stabilization, solar array architectures, combined TT&C and data handling functions, and computer control of subsystem functions. In addition, the design process can select various contemporary and advanced technologies such as the type of solar cell, battery, or propulsion fuel, to name a few. Trade-offs can be tried to improve performance, to reduce weight and power, or for other considerations until the designer is satisfied with the result.

The subsystems are ordered in a logical, interdependent progression, as shown in Figure 8-2. For example, selection of the attitude stabilization technique (spin, 3-axis, gravity gradient) determines the number of solar cells that view the sun at any point in time, which has an important influence on the power subsystem. Selection of the stabilization method will also have an influence on the thermal design. Spin-stabilized satellites evenly expose subsystems to hot and cold space, while 3-axis stabilized satellites can have large thermal gradients caused by specific surfaces exposed to hot or cold conditions for extended periods of time.

Each of the subsystem discussions that follow begin with a review of important mission, payload, and launch vehicle inputs that influence the subsystems calculations.

Attitude Reference and Control (ARC) Subsystem

Depending on the mission, a spacecraft may have varying requirements for pointing accuracies. At some times it may be free to tumble and turn, but at others it may have to pinpoint a discrete location on the earth or deep in space. To do this, the spacecraft must be able to determine its own attitude with respect to some reference, and then to modify this attitude to perform the desired mission. Reference devices include earth horizon sensors, sun sensors, star trackers, or even magnetometers which measure the flux lines of the geomagnetic field. Attitude control devices include tiny thrusters, angular momentum storage wheels, gravity-gradient booms, and electromagnetic torque devices.

A satellite must determine its attitude with respect to some reference (earth, sun, stars, and/or other satellites) and control its attitude to perform the desired mission. Figure 8-4 provides a simplified block diagram of a typical ARC subsystem. The control logic or computer that accepts input from attitude reference sensors is usually based in space but may include ground-based assets in the control loop, directly or as a backup. The mass and configuration of the ARC subsystem are determined by mission and payload inputs that include the final orbital attitude, payload pointing accuracy, and satellite estimated on-orbit mass. These factors influence the types of disturbances that will most perturb the spacecraft body, a few of which are summarized in Figure 8-5. For example, at geosynchronous altitudes, the effects of solar radiation can impart unbalancing torques on the spacecraft body, as shown in Figure 8-6.

It is necessary to select the type of stabilization method that is to be considered. Widely used candidates include:

- 3-axis, zero momentum
- 3-axis, bias momentum
- dual-spin
- spin
- gravity gradient

These stabilization techniques are listed in descending order according to their precise pointing ability.

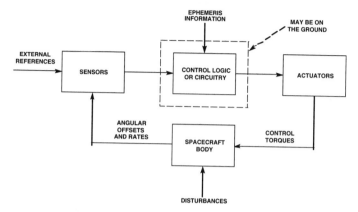

Figure 8-4. Basic attitude reference and control system.

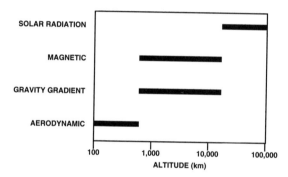

Figure 8-5. Approximate regions of dominance of perturbing forces in space.

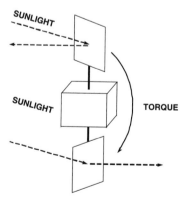

Figure 8-6. Solar radiation propeller torque.

Figure 8-7 provides a rudimentary comparison between dual spin and momentum bias systems. In spin-stabilized satellites, the larger mass satellite bus spins (approximately once per second in geosynchronous orbits) and a smaller mass (communications antennas, remote sensors) is despun so that it can point in an assigned direction (a point on the earth or other planetary body being observed or communicated with). Conversely, the 3-axis stabilization system controls the spacecraft body with one or more internal spinning masses under the direction of the control computer. Figure 8-8 illustrates this even further for a simple body stabilized momentum bias system using a single spinning wheel to control the spacecraft pitch axis while thrusters and/or magnetic torquers provide control forces in roll and yaw.

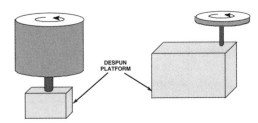

- Moderately complex control
- Moderate accuracy (1.0° to 0.1°)
- Can point to a fixed location

Figure 8-7. Gyroscopic stabilization (dual-spin).

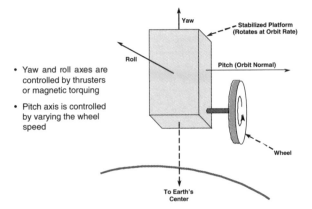

- Yaw and roll axes are controlled by thrusters or magnetic torquing
- Pitch axis is controlled by varying the wheel speed

Figure 8-8. Body-stabilized momentum bias concept.

The ARC subsystem takes its most important cue from the pointing accuracy of the single most demanding payload. This determines the available ARC architectures capable of achieving this accuracy: greater than or equal to 5 degrees for gravity gradient, 1 degree for spin, 0.1 degree for three-axis bias momentum or dual-spin, and less than 0.1 degree for three-axis zero-momentum. Having selected an ARC architecture, the designer matches the selected architecture with the minimum components required for implementation as indicated in the example shown in Table 8-2. The mass for the components, required to implement the selected ARC, is arrived at by a combination of table look-up and calculation. The table look-up mass values are shown in an ARC database such as the one shown in Table 8-3. The contribution to the ARC subsystem mass is simply:

(mass of the component) × (number of components)

As indicated in Table 8-3, the mass for many of the components, such as the reference sensors, is represented by values based on current tech-

Table 8-2
ARC Components Required

Stab Method	Reaction	Momentum	Mag Tor	IRU	Sun	Star	Earth	Mag
1 3-axis, zero-momentum	4	0	2	0	3	0	2	2
2 3-axis, bias-momentum	1	0	2	0	3	0	2	2
3 Dual-spin	0	1	2	0	3	0	2	2
4 Spin	0	0	2	0	3	0	2	2
5 Gravity gradient	0	0	0	0	2	0	0	2

Table 8-3
ARC Components Mass (kg)

S/C Mass (kg)	Wheels	Mag tor	IRU	Earth	Momentum	Sun	Star	Mag
200	2	1	3	2.5	2	0.5	8	1
500	4	1	7	4	4	0.5	8	1
900	6	2	11	5	6	0.5	8	1
1,200	8	3	14	6	8	0.5	8	1
2,000	10	4.5	17	7	10	0.5	8	1

nology which must be continually adjusted to reflect the latest available components. The mass of the actuators that provide control torques to counterbalance disturbances or perform maneuvers are a function of the on-orbit mass of the satellite and include the following components:

- thrusters
- reaction and momentum wheels
- magnetic torquers

The ARC design begins with a default/minimum value for the number of components needed as a function of the stabilization method, as illustrated in Table 8-2. There are two exceptions: Star sensors are only used in the default result if the pointing accuracy requirement is < 0.1 degree, and magnetometers are used only if the spacecraft operating altitude is < 2,000 km.

The designer can make decisions based on special knowledge about the application. For example, the designer may add star mappers to a design that does not require them for pointing to provide, after the fact, pointing knowledge that improves data processing and geophysical application. Another option is to increase or decrease the number of components. Increasing the number of components would build in more redundancy, perhaps extending lifetime.

The selection of an ARC design architecture impacts the power and thermal subsystems. The most prevalent power subsystems make use of solar cell power generation, and the ARC subsystem determines how the solar cells view the sun. Spin-stabilized or gravity gradient ARC designs do not continuously point the entire solar array at the sun as do the 3-axis stabilized ARC designs. This is taken into account in the power system calculation and determines the number of solar cells, of a particular type, that must be used. Correspondingly, the thermal environment of a 3-axis stabilized ARC design is more demanding than that created by spin-stabilized ARC designs, and this will influence the thermal control system mass. As indicated, a spin-stabilized satellite acts like a rotisserie, providing even exposure to hot and cold conditions. A 3-axis stabilized satellite is subject to extreme thermal gradients because particular elements are exposed to hot or cold conditions for extended periods of time.

Power Subsystem

These are the components which provide the power used by the spacecraft to support itself and perform its mission. Additionally, the hardware used to control the distribution of the power is an important element of this system. Solar, nuclear, and other power-generation sources have been used onboard satellites. Examples of power systems used in spacecraft are described in the following paragraphs. Since solar cells are the most common source of spacecraft power, they will be used in the detailed description of the design process.

Solar. There are two types of systems that rely on the sun's radiations to generate power: solar cells and heat exchangers. Solar cells directly convert the sun's photons (light) to electricity by their reaction to the incoming photon flux. Current solar cell technology provides panels capable of generating 140 watts or more of power per square meter of cells. The angle of incidence to the sun is very important and there is a general deterioration of the cells of up to 10% per year due to exposure to the space environment. Heat exchangers use the sun's total radiated energy (the solar constant) to heat a working fluid which then turns a generator to produce electricity. Generated power quantities may be generally higher for heat exchanger systems, but their disadvantage over solar cells is their increased weight and complexity.

Nuclear. There are also two types of nuclear-related power generators which can be used in space: fully critical nuclear reactors and radioisotope thermal generators (RTGs). Nuclear reactors, like solar heat exchangers, use the heat of the nuclear reaction to heat a working fluid which turns a generator to produce electricity. These systems can provide very large quantities of power for missions with high power requirements, such as space-based radars. RTGs also produce heat, but they are noncritical (and inherently safer) and produce lower quantities of electrical power. Both sources are very stable, exhibit long life, can operate far away from solar sources, and can be more compact than solar cells (for maneuvering purposes). Disadvantages are that they are generally very heavy, are radioactive, and are currently in disfavor as a result of adverse public and political reaction to their use.

Chemical. This type of power source includes batteries, with which we are all familiar, and fuel cells in which two elements are recombined with a subsequent release of electrical power such as the oxygen-hydrogen fuel cells currently used aboard the Space Shuttle. All spacecraft employ batteries, at least as storage devices, using the major power-producing system to charge the batteries and subsequently withdrawing power from the storage device during times when power cannot be generated, such as during launch, eclipse times, or malfunctions. Fuel cells represent more power availability than batteries as the quantity depends only on the amount of fuel elements carried. The Space Shuttle cells have an interesting benefit, as the by-product of the power generation is pure, potable water used by the astronauts for drinking. The disadvantage of both systems is their greater weight versus the amount of power produced compared to other methods.

Magnetic. It is also possible to produce power in other ways. One of these could use the process by which a current is produced when a loop of wire is moved through a magnetic field. The earth's geomagnetic field and a spacecraft's motion through it represents a possible future space power source. This potential was demonstrated on a Space Shuttle flight by deploying a payload on the end of a 12.5-mile-long tether, however, it has not been used operationally in space as have the other sources of power discussed above.

Solar Cell Power Systems Design

The power subsystem takes the stabilization method as an input cue from the ARC subsystem to determine the area and mass scaling factors that will be used for power subsystem calculations. Solar panel area and mass scaling factor dependency on the selected stabilization method is approximately:

Stabilization Method	Area/Mass Scaling Factor
3-axis, zero momentum	1
3-axis, bias momentum	1
Spin and dual-spin	3.141593
Gravity gradient	4

Using these values as the foundation for the calculation of the design for the power subsystem, the design is a complex, interactive process that proceeds based on a multitude of mission and payload inputs.

The electrical power subsystem (EPS) supplies electrical power to the satellite, then stores, regulates, and distributes the power as shown in the block diagram of a typical power subsystem depicted in Figure 8-9. Power subsystems designers calculate the subsystem mass by determining the satellite's end-of-life (EOL) power requirement, working backward to determine the satellite's beginning-of-life (BOL) power requirement. BOL power is higher because solar cells degrade over the mission lifetime (influenced by solar cell type, mission altitude, and inclination), and the solar panels must be large enough to guarantee adequate EOL power. When the BOL power requirements are known, the next step is to size the solar array, batteries, power regulation, control, and distribution systems. Solar cell BOL watts/area and watts/mass are selected from a database, such as the example provided by Table 8-4 which lists average values for various solar cell types and array architectures. Solar array sizing takes into account the solar cell type sun incidence angle, battery type, eclipse cycle, power duty cycle, stabilization method, and other related factors. Solar eclipse cycle times versus spacecraft circular orbit altitude are illus-

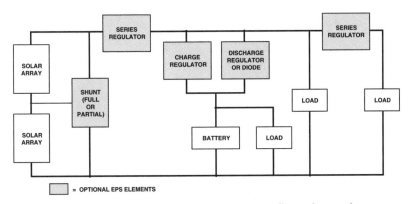

Figure 8-9. Electrical power subsystem configuration options.

Table 8-4
Solar Cell Design Characteristics

	Type	Rigid Array		Flexible Array	
		Power	Mass	Power	Mass
	Name	(watts/m²)	(watts/kg)	(watts/m2)	(watts/kg)
1	SSF Silicon	153	28	153	84
2	Thin Si (APSA)	153	30	153	115
3	GaAs/GE	218	39	218	110
4	Cleft GaAs	223	45	223	163
5	GaAsCulnSe2	272	51	272	150

trated in Table 8-5. Sun incidence angle is selected by the designer to reflect the maximum incidence angle the solar panel will experience. If the solar panels are sun-tracking, the incidence angle is usually set near zero. The final result is the determination of BOL and EOL power and power subsystem mass.

Table 8-5
S/C Eclipse Times Based on Altitude

Circular Orbit Altitude (km)	Solar Eclipse Times (minutes)
200	37.3
250	36.9
300	36.6
350	36.3
400	36.1
450	35.9
500	35.8
600	35.5
700	35.3
800	35.1
900	35
1,000	34.9
1,500	34.8
2,000	35
2,500	35.4
3,000	35.9
3,500	36.4
4,000	36.9
4,500	37.5
5,000	38.1
6,000	39.4
7,000	40.6
8,000	41.8
9,000	43.1
10,000	44.3
15,000	50
20,000	55.2
25,000	60.1
30,000	64.6
35,786	69.4

Several design selections must be made, such as either rigid or flexible arrays and types of solar cells and batteries. Typical battery performance values are shown in Table 8-6. The results of the power subsystem design include EOL and BOL power, determination of solar panel area and mass, number of batteries and their mass, and power control unit, regulator/converter, wiring, and total subsystem mass.

Table 8-6
Spacecraft Battery Characteristics

Battery Type	Specific Energy Density (SMb) (kg/watt hrs)	Allowable Depth of Discharge (%)
1 NiCd	0.040	22
2 NiH$_2$	0.025	60

NiCd: Nickel-Cadmium
NiH$_2$: Nickel-hydrogen

Power Subsystem Design Calculations. As indicated, mission and payload inputs have an important influence on the power design calculation. The final orbital altitude, inclination, and mission lifetime determine the solar degradation rate based on established curves and sets the day/eclipse times that will determine the amount of time the solar cells will be creating power and the amount of time that the payload and supporting systems will rely on battery power. The power required by each element of the payload, the payload duty cycle, and the day/eclipse profile sets the stage for the subsequent system power estimation for the entire satellite. After the baseline power requirements for the satellite are established, the designer makes several important selections that will influence the system mass based on the type of solar array and solar cell desired:

Rigid Array: Derives the mass of the solar array from the solar cell power data base (Table 8-4).

Select the Solar Cell Type: Derives the size of the solar array based on the solar cell database (Table 8-4).

SSF Silicon — Single-sided fixed silicon
Thin Si — Thin silicon
GaAs/Ge — Gallium arsenide/germanium

Cleft GaAs	Gallium arsenide
GaAsCuInSe2	Gallium asenide copper indium selenium
Select the Battery Type:	Drives the specific energy density (mass), and depth-of-discharge (DOD) of the batteries based on the battery database (Table 8-6).
NiCd	Nickel-cadmium
NiH_2	Nickel-hydrogen

Once the input conditions have been satisfied, the power subsystem designer calculates the basic functional values of the system in the following manner:

$$\text{Solar Panel Area (m}^2) = \frac{(\text{BOL Solar Panel Power}) \times (\text{Area Scaling Factor})}{(\text{Solar Cell Watt/sq meter})} \tag{8-1}$$

$$\text{Solar Panel Mass (kg)} = \frac{(\text{BOL Solar Panel Power}) \times (\text{Mass Scaling Factor})}{(\text{Solar Cell Kg/sq meter})} \tag{8-2}$$

The minimum number and mass of the batteries is calculated based on the energy storage needs. The number of batteries may be increased for redundancy.

$$\text{Min. Number of Batteries} = \frac{(\text{Total WH Stored})}{([\text{Bus Voltage}] \times [\text{AH/Batt}])} \tag{8-3}$$

$$\begin{aligned}\text{Battery Mass (kg)} = &\ (\text{Bus Voltage}) \times (\text{AH/Batt}) \times \\ &\ (\text{Num of Batts}) \times (\text{Batt Mass to Power Density})\end{aligned} \tag{8-4}$$

The mass of the power control and conditioning systems is based on a percentage of the power produced.

$$\text{Power Control Unit Mass (kg)} = 0.02 \times (\text{Day Power}) \times 2 \tag{8-5}$$

Regulators/Converters Mass (kg) = 0.025 × (Day Power) × 2   (8-6)

The mass of the wiring is based on a percentage of the overall spacecraft mass.

Wiring Mass (kg) = 0.04 × (Estimated S/C Dry Mass)   (8-7)

These values define the output of the power subsystem which constitutes the largest mass subsystem and one of the more complex elements in the satellite design.

Thermal Subsystem

The thermal control system keeps the temperature of the spacecraft within specified ranges. Too much built-up temperature may affect electronic systems, causing failures, and low temperatures may freeze up movable systems or cause fuel lines to freeze and burst. The control system must be able to react throughout the expected environmental and operational configurations experienced. Low-earth orbiting spacecraft continuously cycle in and out of the sun's radiations, absorbing and radiating energy in phases. Geostationary satellites may be exposed to (or eclipsed from) solar radiation continuously for weeks and months, and sun-synchronous satellites may be continuously exposed. Internally, a spacecraft may generate large quantities of heat during times of peak operation, and little during periods of dormancy.

Thermal control devices fall into two categories: passive and active. Passive devices simply shield, insulate, or change their thermal characteristics depending on the existing temperature of the satellite. The external coating of a spacecraft determines the craft's *absorptivity* (how much external energy is absorbed) and *emissivity* (how much internal thermal energy is radiated into space) characteristics to control temperature. The space shuttle rotates in a maneuver called *rotisserie* to alternately expose the black underside and the silver-coated open cargo bay doors for the purpose of thermal control. Many spacecraft are wrapped in thermal blankets to retain internal heat, and some spacecraft are equipped with passively heat-activated louvers which open or close to expose different external surfaces to radiate or retain internal heat.

Active devices usually involve some sort of working fluid to carry heat from one location within a spacecraft to another to either increase or decrease internal or even local temperatures. Refrigeration devices, heat pumps, and heat pipes are examples of active spacecraft thermal control systems. These devices have the ability to more precisely control spacecraft temperatures, but their disadvantage is their increased weight and complexity. Special thermal systems, such as cryogenic coolers, are usually accounted for as unique payload support systems. Thermal subsystem mass is usually estimated as a percentage of spacecraft dry mass, typically 2–5% for passive systems and 4–8% for active systems.

Orbital Maintenance Subsystem

This subsystem design determines the amount of propellant needed to keep the satellite in the desired orbit over the specified mission lifetime. Atmospheric density causes drag and is the primary force that will cause many spacecraft to lose altitude. Atmospheric density is a function of the solar environment, which can vary widely as a function of the expected solar cycle activity, and satellite altitude, as shown in Table 8-7. The final orbit altitude and mission timeline is used to select the appropriate value of atmospheric density. The satellite ballistic cross-section, which is the cross-sectional area of the satellite in the direction of flight, determines the amount of drag. A mission duration Δv is calculated to determine the total propellant needed to maintain the orbit. The amount of propellant

Table 8-7
Atmospheric Density

Altitude (km)	Expected Solar Cycle Activity (kg/m^2)		
	Low	Average	High
100	9.8E-09	9.8E-09	9.8E-09
200	1.95E-10	2.7E-10	3.45E-10
300	8E-12	2.23E-11	3.65E-11
400	1.04E-12	4.77E-12	8.5E-12
500	1.21E-13	1.05E-12	1.98E-12
600	2.45E-14	3.12D-13	6E-13
700	7.1E-15	9.3E-14	1.8E-13
800	3.4E-15	3.4E-14	6.4E-14
900	2E-15	1.4E-14	2.5E-14
1,000	1.3E-15	6.7E-15	1.2E-14

needed is influenced by the specific impulse of the propellant (Isp). The higher the specific impulse, the greater the energy, which results in less propellant needed for on-orbit maintenance. Typical specific impulse values are shown in Table 8-8. Both monopropellants such as N_2H_4 (hydrazine), and bipropellants such as N_2H_4/N_2O_4 (hydrazine/nitrogen tetroxide) are used for orbital maintenance. Depending on spacecraft operational altitude, atmospheric drag may be compensated by an electrostatic system which produces a small amount of thrust over long periods of time.

The details of an orbital maintenance subsystem design depend upon the specifics of the satellite orbit (LEO, SSO, or GEO). For GEO missions, north/south and east/west station-keeping requirements must be considered due to perturbations produced by sun and moon gravitational effects as well as nonuniformity of the earth's gravitational field. The output of the design process is the mass of the orbital maintenance fuel and associated tankage, plumbing, and thrusters. Tankage and plumbing may be shared with the *propulsion* subsystem; however, in many instances, thruster sizing and placement is unique to the orbital maintenance subsystem.

Propulsion Subsystem

The satellite propulsion subsystem is used to take the spacecraft from a parking orbit to a final orbit. The spacecraft design may choose a "standalone" stage for this task. In most cases, modern spacecraft will use a com-

Table 8-8
Thrust System Specific Impulse

Type	Propellant	Isp
Liquid	Monopropellant 220	220
Liquid	Bipropellant 350	350
Liquid	Bipropellant 425	425
Liquid	Bipropellant 430	430
Liquid	Bipropellant 450	450
Electrostatic	Resistojet 400	400
Electrostatic	Resistojet 700	700
Electrostatic	Arcjet 500	500
Electrostatic	Arcjet 1,000	1,000
Electrostatic	Arcjet 1,500	1,500

mon or integrated system for orbital maintenance and propulsion and not have a separate perigee or apogee kick motor. Therefore, the specific impulse for both the satellite propulsion subsystem and the orbital maintenance subsystem have the same value, and the systems share some common tankage and plumbing. Thruster placement, design, and specific fuel mass may be unique for either orbital maintenance (station-keeping) or orbital translation (propulsion). Since many spacecraft combine these functions, a more detailed description of a composite system is explained in this context.

Propulsion and Orbital Maintenance As a Combined Subsystem

In most cases, onboard propulsion systems must be included to allow the spacecraft to modify its orbit (if necessary), maintain the orbit despite the perturbing forces it may encounter (orbital maintenance), and sometimes, as an adjunct, to control the desired attitude of the spacecraft. To accomplish this, the spacecraft may have a relatively large rocket for orbit modification, smaller rockets for station-keeping, and tiny thrusters for attitude control. Associated with each of these are fuel storage, delivery, and control devices.

The propulsion and orbital maintenance subsystems are listed separately based on a design heritage that recognized that many traditional, modular satellite designs had completely separate propulsion and orbital maintenance systems. For example, some designs used solid-fueled propulsion systems to translate from the parking orbit (achieved by the launch vehicle) to the final orbit and a separate liquid-fueled orbital maintenance system for station-keeping during the life of the satellite. However, many contemporary designs combine the propulsion and orbital maintenance systems using a common fuel and storage system and specific and separate propulsion and orbital maintenance thruster designs that are unique to the function of orbital translation and orbital maintenance, respectively. The fuel mass is specific for the respective propulsion and orbital maintenance categories where they accrue, and will be explained in that way in the following paragraphs.

The propulsion and orbital maintenance calculations are both based on the Δv or change in velocity (vector) required to perform either an orbit translation or orbital maintenance. Both calculations are based on the equations associated with orbital mechanics that determine the Δv required for making a change from one orbit to another. In the case of the propulsion subsystem, the change is associated with going from the parking orbit to the final orbit. In the case of the orbital maintenance system, the designer calculates the atmospheric drag on the satellite associated with the precise mission lifetime (calendar years), which invokes the expected spacecraft environment during this lifetime expressed in terms of atmospheric density as a function of orbit altitude. Atmospheric drag acts in the opposite direction of the orbital velocity vector and removes energy from the orbit. This reduction of energy causes the orbital radius to get smaller, leading to further increases in drag. Acceleration due to drag on a satellite is a function of the atmospheric density, the satellite cross-sectional area in the direction of flight (ballistic cross-section), and the satellite mass (in this case the on-orbit mass). The drag is calculated using the expected environment and the satellite ballistic cross-section. Various shapes and densities exist from the distributed modular design architectures to the dense, highly integrated SMALLSAT designs. This diversity in density and satellite shape results in an individual determination of the ballistic cross-section of a particular satellite. These are all known quantities and are used, in turn, to calculate the $\Delta \mathbf{v}$ required to restore the satellite to the prescribed orbit. The remaining quantity required for this calculation is the $\mathbf{I_{sp}}$ or specific impulse of the fuel that will be used to perform this maneuver. The specific impulse is a measure of the energy content of the propellant used and how efficiently it is converted into thrust, or how thrust is produced per time rate of change of propellant molecular weight. The $\mathbf{I_{sp}}$ of the system to be used by the propulsion and/or orbital maintenance subsystems can be selected by the designer from a database such as the one shown in Table 8-8.

The orbit modification propulsion subsystem has to provide significantly more $\Delta \mathbf{v}$ than the orbital maintenance subsystem. The propulsion subsystem must provide the thrust to take the spacecraft from the parking orbit to the final orbit. The required performance of the system is expressed as a required $\Delta \mathbf{v}$ which is calculated using the principals of orbital mechanics. Usually only the magnitude of the velocity vector is changed (the condition when the parking orbit inclination or plane is the same as the final orbit), but sometimes a plane change is required (the situation for the positioning of most geosynchronous communications satellites). A design must calculate the $\Delta \mathbf{v}$ required for either one of these situations and, using this and the initial mass of the satellite ($\mathbf{m_o}$) and the $\mathbf{I_{sp}}$ of the selected fuel, proceeds to determine the fuel mass ($\mathbf{m_p}$) required to perform the maneuver:

$$m_p = m_o \left[1 - e^{-(\Delta v / I_{sp} g)} \right] \tag{8-8}$$

where $\mathbf{g}$ is simply the gravitational acceleration.

Thruster and tankage mass are calculated using accepted thruster component and tankage values as a function of the total fuel mass, the number of thrusters required, or other factors. Typical values are 10% of the required fuel mass. After the mission and payload input has been completed, the designer estimates a spacecraft dry weight (no fuel) and then, using the above method and fuel calculations, makes an estimate of the spacecraft wet (parking orbit) and on-orbit (final orbit) mass estimates. These values are then used to make new estimates for the propulsion and orbital maintenance subsystems. The process is repeated iteratively until an accessible estimate is obtained.

Data Handling Subsystem

Most spacecraft are engaged in the collection or transfer of some form of information, even if the specific mission performed by the spacecraft is not involved with communications. The data handling subsystem collects data from the payload and sends the data to the ground. The collected data may be transmitted as it is collected or recorded for later transmission. Data transmission is either direct to a ground station and/or to a ground station via a communication relay satellite, such as the *tracking and data relay satellite* (TDRSS). The designer must select the downlink method and frequencies and indicate if onboard recorders are used. A design will calculate the downlink rate (Kbps) and present the downlink options for users to select from. There are several communication bands normally used to send the data to the ground, such as:

S-Band	(2–4 GHz)	Low data rate (could be shared with TT&C)
X-Band	(8–10 GHz)	Payload specific high data rate
K-Band	(13–15 GHz)	Payload specific high data rate

After the communication band(s) are selected, the designer calculates the subsystem envelope, mass, and power requirements using a database of transponder, antenna, and recorder information. Conventional designs assume that multiple transponders, antennas, and recorders are always used for redundancy. The result of the data handling subsystem analysis is the determination of subsystem mass and power.

TT&C Subsystem

Associated with most spacecraft is the telemetry signal which contains information on the status of the spacecraft and its systems and through which a controlling ground station can modify the spacecraft's instructions (commands) for future actions. Modern sophisticated spacecraft are equipped with solid-state control devices on which mission logical instructions can be placed and through which the spacecraft will perform those instructions autonomously. As part of this overall spacecraft coordination, the controller may interrogate its separate systems for their status and perform corrections in cases of malfunction or failure.

The *telemetry, tracking, and control* (TT&C) subsystem usually has a low data rate (1–2 Kbps) and can do double duty serving as a downlink for low-rate payload data. If the composite payload data rate is low, it can be accommodated by the TT&C subsystem. If the composite data rate is high, a payload specific channel is required in the data handling subsystem.

Like the data handling subsystem, a TT&C design calculates the subsystem mass and power requirements from available systems assuming two transponders and omnidirectional antennas for redundancy.

Data Handling and TT&C As a Combined Subsystem

The data handling and telemetry, tracking, and command (TT&C) subsystems, while separate in the subsystems listing, are associated in much the same way as the propulsion and orbital maintenance subsystems. To conserve weight and space, a design may take advantage of the shared use of telecommunications resources. If an S-band TT&C link can also be shared to collect the payload information being collected, then satellite complexity and cost is reduced.

The data handling requirements are based on the payload input values of data rate, duty cycle, and whether or not onboard recording is specified. The data rate will determine whether a low-frequency or high-frequency channel is required for transmission (see the information bandwidth description in Chapter 5) and, hence, whether the low-frequency (S-band) TT&C channel can be shared for remote payload data collection. If the data rates warrant, a wider bandwidth channel (X-band or even K-band) will be added as a separate data handling subsystem. A designer could amend this by taking advantage of a low-percentage duty cycle (instrument on time of a few minutes per orbit) and use a recorder playback at a lower rate to return to a shared data handling TT&C design. If the payload input has onboard recording as a requirement, the designer can take this into account in determining the mass of the data handling subsystem. The designer is still responsible for selecting the type of recorder desired (magnetic tape or solid state).

Once the component suite is complete, the designer tallies the mass and power requirements for the combined data handling and TT&C subsystem using contemporary component values. The elements include transponder mass, hemispheric and directive parabolic antenna mass, and wiring mass. Typical values for a combined TT&C and data handling subsystem are 2–6% of spacecraft dry mass.

Onboard Computer Subsystem

The onboard computer subsystem asks the designer to specify what function the computer will be used for. The functions include:

- attitude control
- navigation control
- orbital maintenance
- command processing
- telemetry processing
- payload management
- power management

The onboard computer may control functions such as power and data management, recovery from failure, and others. The onboard computer system mass and power are estimated as a function of computer memory storage size (K-words) and throughput (Kbps). The designer indicates which subsystems are communicating with the onboard computer. Based on this selection, a design estimates the storage word size and the throughput and establishes the subsystem mass and power.

The designer can also specify whether the satellite is designed with simplified or complex "stand-alone" autonomy. Complex autonomy provides significantly more computer capability because it is associated with systems that are required to make important operational and emergency decisions without benefit of ground analysis or control and may operate for several months without any ground interaction whatsoever. The default design decision is to use full redundancy in this important design area. A designer can change this decision if application goals, i.e., special limited lifetime missions, so warrant. In many designs, mass and power requirements for this subsystem are accounted for with the TT&C and data handling subsystems.

Structure Subsystem

Spacecraft structural considerations must deal with the spacecraft in its many physical configurations. It must be able to support itself under the heavy load of the earth's gravitational pull, survive the violent process of launch and positioning, be able to deploy systems into an operational configuration from the folded, compact configuration of launch, and finally, to maintain spacecraft integrity during station-keeping and attitude changes.

The mass of the satellite structure may be estimated as a percentage of the satellite dry mass, typically around 20%. However, the structure mass is also influenced by payload and payload support dynamics, such as scanning antennas and magnetic tape recorder transport, and operating power levels and profiles.

An alternate estimation of the mass of the structure subsystem is similar in calculation to the previously described thermal subsystem calculation, with some important exceptions based on uncompensated momentum mass imbalance created by scanning payloads. The calculation is iterative beginning with the default values obtained for the mass of each of the spacecraft subsystems, recalculating at the end of the design process. An example calculation might proceed as follows:

For each element of the payload the mass multiplier is 0.1 if the payload is passive and 0.15 if the payload is active. The payload mass for this calculation is the total mass or the sum of the electronics and sensor mass. If the payload has a scanning antenna that is uncompensated, a compensating mass is added to the structure weight.

For the ARC and power subsystems structure, the mass multiplier is 0.1 if all of the payloads are passive and 0.15 if any of the payloads are active. The mass multiplier for each of the remaining spacecraft subsystems is 0.1.

Each of these contributors are summed to produce the structure subsystems mass which is allocated to the contributing elements. Results using this type of an approach are consistent with representative designs.

Design Iteration

Since the determination of the mass of a spacecraft subsystem depends upon the mass of each of the other subsystems (for example, ARC mass will depend upon payload requirements as well as the mass of all other subsystems), a design must employ an iterative approach for establishing subsystem masses. In other words, as each subsystem is designed, all known masses are taken into account. Upon completion of the design of all subsystems, an iteration process will be initiated whereby the mass of each subsystem is recalculated, taking into account knowledge of the masses of all other subsystems. This cycle is repeated until the mass computations stabilize from one iteration to the next. By this process, all subsystems take into account the mass of each of the other subsystems in their mass computations.

Table 8-9 provides an overview of the characteristics of some historic and contemporary spacecraft systems. The mission values that drive each design are listed along with some of the top-level characteristics of the satellite bus subsystems. The table provides a perspective about the relative size of subsystem mass values.

GROUND SUPPORT SYSTEMS

Another major component in a space system is the ground support system. This consists of elements that support the establishment and operation of the spacecraft. Ground systems include the launch, tracking, status monitoring, command, and data collection services required to establish and maintain the spacecraft during its lifetime in orbit.

Once the spacecraft is constructed, the support personnel and systems integrate the spacecraft with the launch systems and monitor it before, during, and after the launch, which includes positioning and check-out.

Table 8-9
Comparative Overview of Some Typical Spacecraft Systems

		GEOSAT FO	LEWIS	CLARK	TOPEX	RADAR-SAT	LAND-SAT 6	TIROS-N	ERS-1
	Altitude (km)	800	517	475	1,136	800	705	870	800
	Inclination (deg)	66	97.4	97.3	98.8	98.8	98.2	98.8	98.8
	Launch Year	1996	1996	1996	1992	1995	1993	1992	1992
	Life (yrs)	3	5	5	3	3	5	3	3
	Satellite Dry Mass (kg)	286	387	278	2,190	3,100	1,750	1,030	2,160
	Ballistic Area (m2)	1.2	1.5	1.2	6.5	12.8	5.8	4.1	13
	Stabilization	3-AXIS	3-AXIS	3-AXIS	3-AXIS	3-AXIS	3-AXIS	3-AXIS	3-AXIS
ARC	Sun Sensor	YES	YES	YES	YES	YES	YES	YES	YES
	No. of Star sensors	0	2	2	2	0	2	0	2
	Mass (kg)	34	38	40	205	156	130	77	156
Power	End of Life (W)	738	740	581	2,140	2,700	1,900	1,500	3,000
	Mass (kg)	113	140	104	550	640	380	200	650
Thermal	Mass (kg)	7	12	8	54	93	29	46	65
Data/TTC	Mass (kg)	45	42	45	177	216	325	66	87
	Power (W)	66	60	66	0	0	0	0	0
Structure	Mass (kg)	30	44	38	480	636	250	217	488

After the spacecraft is declared operational, the support personnel and systems continue to monitor it for health and operability. Meanwhile, the operations support system has begun to utilize the spacecraft and support this use. This includes ground transmit and receiving stations and data collection, reduction, and dissemination systems.

Spacecraft Design and Construction

Depending on the originator of the spacecraft idea, design and construction may occur in many different locations. Some experimental systems may be the product of university-associated laboratories, such as the Johns Hopkins Applied Physics Lab (APL), and the California Institute of Technology's Jet Propulsion Lab (JPL). The National Aeronautics and Space Administration (NASA) may create some of its own spacecraft at any of its diverse locations, such as the Goddard Space Flight Center (GSFC) or Kennedy Space Center (KSC). Preliminary military space system design may occur at the Air Force's Space Division or the Navy's Space Systems Activity (SPAWAR). But, in most cases, operational spacecraft will be designed and built by industrial facilities based on *requests for proposals* (RFPs) presented by the activity desiring the spacecraft to be built.

Launch Sites and Vehicles

After a spacecraft has been built and tested, it is transported to the launch site to be mated with the launch vehicle for delivery into space. In the United States, the two main launch sites are the eastern test range (ETR) at the Kennedy Space Center in Florida and the western test range (WTR) at Vandenberg Air Force Base in California. Airborne launches from diverse locations have also become available.

Table 8-10 lists the total launch vehicle launch weight, lift-off thrust, and maximum payload weight to a typical low altitude orbit for some select launch vehicles. All numbers are approximate.

As indicated earlier, launch vehicles are designed simply to deliver their payloads into space and are usually not capable of reaching the final orbits from which many systems operate. Therefore, upper stages may be used to position the spacecraft into (or close to) its operational orbit. As mentioned before, most spacecraft have an associated propulsion system which is used for final positioning and possible repositioning, but the

Table 8-10
United States Launch Vehicles*

	Launch Weight (lbs)	Lift-off Thrust (lbs)	Low-orbit Payload (lbs)
Delta	426,000	631,100	6,000
Atlas	292,600	378,000	11,200
Titan	1,519,600	2,920,000	34,050
Shuttle	4,500,000	6,425,000	65,000

Typical performances of the most common U.S. launch vehicles are given.

upper stage is a separate rocket which is usually separated from the space-craft after use.

An important point is that if an upper stage is required, it usually sub-tracts from the available weight and space for the payload in the launch vehicle. (Because launch vehicles are multistage rockets, some of the final stages may be capable of delivering the payload into the desired orbit with-out the use of an additional upper stage.) Some common upper stages are the liquid-fueled Centaur (used to send the Viking and Voyager probes away from the earth), and the solid-fueled Payload Assist Module (PAM) and Inertial Upper Stage (IUS), both capable of use from the space shuttle.

Ground Sites

Ground sites consist of the spacecraft control facilities, which monitor and task the system in orbit, and the receiving sites to which the spacecraft may transmit information for dissemination and use. These are many and varied and depend on the operator of the spacecraft. At the receiving sites, infor-mation is collected and put into a form suitable for dissemination to the users.

On the ground is also the most important component of a complete space system: the system user. In many cases, the user drives the system design and later operation, as indicated in Table 8-11 which illustrates the generic personality of communication, navigation, remote sensing, and planetary systems. In many instances, the user only accesses the informa-tion for whatever purpose the spacecraft was designed in the first place. The task force commander at sea only knows that he has instant access to his units, via the satellite communications link, and knows his and others' positions accurately due to the availability of satellite navigation. Weath-er information in his area of operations is provided by weather processing

Table 8-11
Satellite Personality of Generic Systems

	Communication	Navigation	Remote Sensing	Planetary
Orbit	Geosync, Molniya, LEO	High Earth Circular	Low Earth Circular High Eccentric Geosync	All Types
Power	Solar	Solar	Solar	Nuclear, Chem, Solar
Attitude	Spinning, 3-Axis	3-Axis	3-Axis, Grav Grad	3-Axis
Data	High	Low	High + Proc	Low
Size	Moderate to Large 1,000–5,000 kg	Small 500 kg	Moderate to Large 1,000–5,000 kg	Small to Moderate 500–5,000 kg
Mission	Data Relay	X-Mit Position	Collect Info	Various
Lifetime	8–12 Years	8–12 Years	3–5 Years	Years
Cost	$50–80M	$5–30M	$50–500M	$0.1–5B
Complexity	Moderate	Low	High	New Horizons
Lead Time	2 Years	2 Years	4 Years	5 Years

and dissemination ground sites that allow him to best utilize his airborne, surface, and subsurface assets. Every time you pick up the phone and make a long-distance call or watch cable TV at home, you become a user of spacecraft systems.

SPACE SYSTEM EVOLUTION

The space system design process has evolved over the nearly four decades in which it has been conducted. Table 8-12 provides a historical perspective of the evolution of spacecraft systems over this period. Early space systems were small (due to the existing launch vehicle constraints) single-purpose systems. As launch capabilities increased, so did the size and complexity of spacecraft designs, and billion-dollar, multimission spacecraft have been built and launched. Lessons were learned with these systems when small failures of single systems either decreased the effectiveness of the spacecraft or rendered it totally useless.

Modern spacecraft design uses both multiple-use, modular construction spacecraft designs with interchangeable systems using more common (and inexpensive) off-the-shelf equipment and the recent resurgence of smaller, single-purpose, highly integrated spacecraft that bear the mantle of SMALLSATs and LIGHTSATs. More and more spacecraft are being

Table 8-12
Evolution of Satellite Systems

	1960s/70s	1970s/80s	1990s/2000s	
SPACE SEGMENT	Small Integrated design Expendable	Large Modular Expendable	Very large, multiple-use Modular, expandable Long life with maintenance and modification	Small, single use Short life
MISSION	Navigation, communication, passive remote sensing Single missions	Navigation, communications, active remote sensing Integrated missions	Many missions using one platform (e.g., space shuttle, space station)	Single missions
CONTROL SEGMENT	Ground-based Low data rate Frequent blind periods Spotty tracking ground stations	Ground- and space-based High data rate Selective ground network and space relay Nearly continuous coverage Good tracking, varying with orbit	Autonomous and space-based Onboard processing, evaluation, decision-making Continuous coverage Excellent continuous tracking	

designed with the capability for in-space repair or retrieval for refurbishment and reuse. The future will see major space systems actually constructed and tested in space before deployment and operation.

REFERENCES/ADDITIONAL READING

Pisacane, V. and Moore, R., *Fundamentals of Space Systems.* New York: Oxford University Press, 1994.

Wertz, J. and Larson, W., *Space Mission Analysis and Design.* Boston: Kluwer Academic Publishers, 1991.

Griffin, M. and French, J., *Space Vehicle Design.* Washington, D.C.: American Institute of Aeronautics and Astronautics, 1991.

Kaplan, M., *Modern Spacecraft Dynamics and Control.* New York: John Wiley & Sons, 1976.

EXERCISES

1. A geostationary communications satellite requires 1,000 watts of power to serve its payload and spacecraft bus subsystems. Calculate the number of SSF Silicon solar cells required for a 10-year design life for a spin-stabilized spacecraft. Compare this result to a 3-axis stabilized design.

2. Estimate the power subsystem mass for each of the spacecraft designs in Exercise 1 above.

3. The on-orbit mass of a satellite is 1,000 kg. The satellite begins life in a 66° inclination, 500 km altitude parking orbit achieved by the launch vehicle. Calculate the mass of propellant required to place the satellite into its final orbit of 99° inclination and 1,000 km altitude using bipropellant 350.

4. A satellite with a ballistic cross-section of 4 m² is in an 800 km altitude, 108° inclination orbit where it is expected to perform for 5 years. Calculate the fuel mass required using monopropellant 220. Compare this result to the case in which the satellite's orbital altitude is 300 km.

CHAPTER 9

Spacecraft Design

In the complex world of aerospace vehicle design, it is no longer possible for a single designer to individually create a vehicle which can optimally and capably perform the many functions that modern expectations place on such systems. The word "system" is not used lightly here, as more and more often, modern-day systems are complex assemblages of many different parts. With this in mind, the engineering definition of a "system" may be introduced:

A *system* is a set of *elements* organized to perform a set of *designated functions* in order to achieve *specified results*. This definition itself introduces additional terms which, in turn, must also be defined:

An *element* of a system is an identifiable component of the whole that performs some identifiable function. Each element performs particular *designated functions* which represent the defined actions, operations, or use of the element. These well-defined actions assist the system in achieving its *specified results* which are simply the stated objectives or purpose of the system.

Thus defined, it is easy to see that an element may be a system in its own right. More than just individual pieces of a particular mechanism, system elements may also include personnel, facilities, information, and many other factors that are not necessarily physical parts of the mechanism under design.

In viewing the system as a set of many diverse components, it can easily be seen that a single designer, though competent in one or even a few fields, may be lacking in some knowledge that may not allow him or her to compose the optimum method for achieving the particular objective. Designs of large-scale, complex modern-day systems is a multidisciplinary task calling on the expertise of many toward the one objective.

THE SYSTEMS APPROACH

In order to coordinate the efforts of those contributing to the task, an organized method must be followed to ensure success. The *systems approach* described herein, or something similar, is the methodology used to approach the problems encountered by NASA and the aerospace industries in meeting the needs of the space programs, by AT&T in meeting the communications needs of millions, and even by urban planners considering public transportation, pollution, health care, and other needs.

From the above discussion, it should be clear that the prerequisite for considering the design of a space-based system is that an appropriate mission need exists. A suitable commercial, military, civil, or scientific need must be well established to justify the time, effort, and sizable expense required to warrant the development of a space-based system. The need is often defined by individuals or groups of potential systems users scattered all over the world. Sources of demands for satellites originate from military or civil government agencies, scientific organizations, and commercial service industries. Some space systems may be sovereign in nature, providing information and services to a select but important sector of users. In some cases, international organizations, both scientific and commercial, form shared programs with many entities providing funds and systems elements. Advisory committees, such as the National Academy of Sciences and the National Space Council, and universities provide guidance and play important roles in defining needs for manned and unmanned satellites. Whatever the source of the initial concept for the mission or the selection of the specific implementation approach of the system, further development usually follows a fairly common, relatively standardized and structured approach, referred to as the *systems approach.*

As shown in Figure 9-1, the systems approach is a two-dimensional method which has the *design process* along one axis and the systems *life cycle* along the other. When placed together, the result is a matrix which delineates a methodical system design process. Each step in this process is explained in the following sections.

System Life Cycle

The system life cycle begins with formalization of the mission need and terminates when the system is retired or fails. The life cycle may be originated as a result of a new need, new technology, or as an iteration of a previous system whose life cycle is nearing completion due to obsolescence. As indicated in Figure 9-1, the system life cycle has three distinct periods: *planning, acquisition,* and *use.* Each of these periods is divided

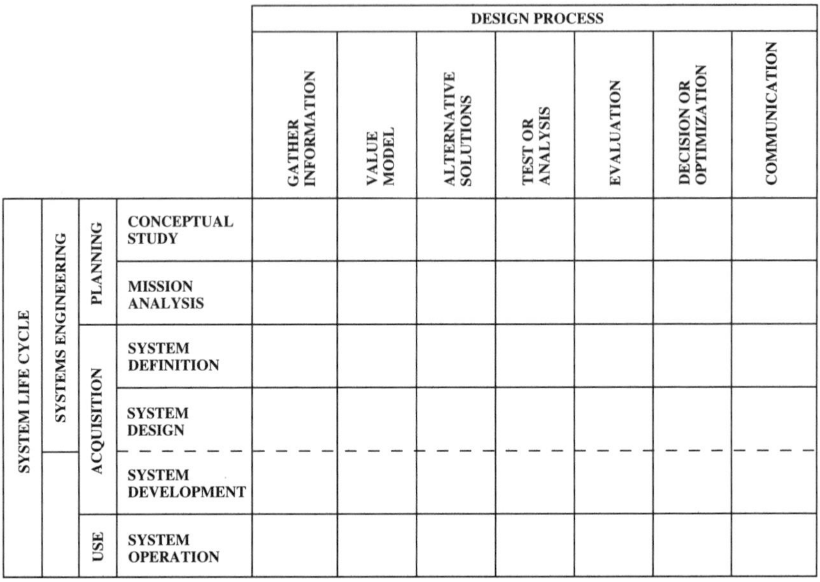

			DESIGN PROCESS					
			GATHER INFORMATION	VALUE MODEL	ALTERNATIVE SOLUTIONS	TEST OR ANALYSIS	EVALUATION	DECISION OR OPTIMIZATION	COMMUNICATION
SYSTEM LIFE CYCLE	SYSTEMS ENGINEERING	PLANNING	CONCEPTUAL STUDY						
			MISSION ANALYSIS						
		ACQUISITION	SYSTEM DEFINITION						
			SYSTEM DESIGN						
			SYSTEM DEVELOPMENT						
	USE		SYSTEM OPERATION						

Figure 9-1. Systems approach. A structured approach is used in the design of complex systems such as spacecraft.

into a number of *phases* through which the designer conducts the distinct steps of the systems approach.

The initial period in the life cycle is the *planning period*. During this period, the need for the system is established, system concepts are formulated, and their feasibility and worth are established. The output of the planning period is an identification of a system concept and a set of system requirements. The *acquisition period* includes those steps necessary to design, test, and evaluate the system, and to produce and install it. The *use period* consists of those activities and resources required to operate, support and maintain the system.

The phases associated with the planning period consist of the *conceptual study* phase and the *mission analysis* phase. The acquisition period starts with the *system definition* phase and ends with the system development phase which includes launch and initiation of operation. Within these two periods, the activities associated with *systems engineering* are defined. It is during these periods that maximum benefit may be realized in following the design process of the systems approach. As shown in Figure 9-1, these are mainly applicable to the planning phases and the phases concerned with design of the system prior to production. The planning, acquisition, and use periods represent *time*-based periods in a system's life cycle.

System Design Process

While the system life cycle identifies the different periods and phases in the design of a complex system, the *design process* defines the fundamental sequence of activities used for making design decisions (trade-offs) during each phase. It is an iterative (repeated) method that results in a successively more detailed and optimized solution to the problem at hand. The process is continued repeatedly until either an acceptable solution is reached or a decision is made to regress to a previous design phase (or even back to square one!). The process is broken into a number of *stages* which are also indicated in Figure 9-1. These stages are discussed in more detail in the following section, after which the phases in the system life cycle will be presented in detail.

Design Process Stages

Gather Information. In the first stage, the designer (or group) gathers all the information that may be pertinent to the problem. This includes the inputs to the particular design phase (defined later), as well as overall system constraints and purposes, which may be useful in establishing criteria to which the inputs will be compared and optimized.

Value Model. The next stage uses the information gathered in the first stage to formulate a function (model) that indicates the *desired* performance of the system or element with respect to all pertinent considerations (such as size, cost, and effectiveness). It is to this model that the evaluation functions of all the possible approaches will be compared. In general there are three important areas to consider:

- *Effectiveness:* ability to perform the stated objectives. This mainly deals with the physical capabilities of the solution.
- *Cost:* resources required by the solution.
- *Time:* when the solution is required or how long it may take to produce the particular solution.

Many other considerations may be involved such as risk, politics, and others directly associated with the particular need being addressed. These considerations are not to be isolated from one another, and each may contain a number of factors within itself. Each factor must be identified and the relative importance of each specified in the evaluation function by use of weighting factors. An example may be given for the construction of a road for which we would like to minimize the time between destinations and maximize the efficiency (traffic control), but, of course, minimize the cost. The parameter may look like:

$$\text{ROAD}_{(\text{time} \times w_1, \text{ efficiency} \times w_2, \text{ cost} \times w_3)} \qquad (9\text{-}1)$$

where time, efficiency, and cost represent the important variables of the parameter and w_1, w_2, and w_3 represent the weighting factors assigned to each variable. These variables must all be considered, but some may be more important than others and will be given higher weighting factors in the actual parameter statement indicating their priority. Some of these variables may also conflict with the desires of some of the others. For instance, minimizing the travel time on our road by cutting a hole through a mountain may not fit into the cost restraints with which we also must deal.

A considerable amount of effort must be put toward this stage of the design process since it directly affects the outcome of evaluation of possible solutions to the problem.

Alternative Solutions. Having decided on the desired performance during the last stage, the designer may now use his knowledge and analytical skills to synthesize solutions to the problem. Formulation of solutions prior to the last stage should be avoided to prevent biased ideas. This is the fun part of systems engineering where the designer may dream up exotic solutions to the stated objectives and have them compared to the more conventional solutions that will also be brought forward. (Who knows if the exotic solution won't out-perform the conventional solutions when the evaluation stage is reached!)

Test or Analysis. Before a good evaluation can be performed, as much information as possible on each solution must be obtained. In the test or analysis stage, each solution is tested (using models or simulations), if possible. Alternately, its performance in each of the value model factors may be estimated. This will result in a performance that can be compared to the desired performance model formulated earlier.

Evaluation. The results of the test or analysis stage for each solution are compared to the desired performance function *as well as to the results for* *each of the solutions considered.* It must be realized that none of the possible solutions may meet all the criteria specified in the desired performance function completely or to the level desired, so comparison of the solutions obtained may be beneficial.

Decision/Optimization. At this point, the designer must make a decision. This process is simplified in the systems approach by the fact that only four courses of action are possible:

- One (or more) of the solutions meets an acceptable level of performance. The solution which gives the *best* level is usually the one that will be passed on as the solution to the next phase of the design process, though more than one may be passed on as alternative solutions as well.
- None of the solutions represents an acceptable solution, but one of the alternatives exhibits promise if reevaluated or refined within the same phase of the design. This may involve refining the solution itself in terms of its configuration or analysis, or may require redefinition of the evaluation function itself. This process of feeding back to a stage within the same step of the design process is known as *optimization.*
- None of the solutions are acceptable and a return to an earlier stage is required.
- There is no solution and the program should be dropped.

Communication. The last stage of the design process involves communication of the results. This must be in a form suitable for use by the follow-on activities and may involve reports, drawings, specifications, or models. This represents the end of the design process for a particular step, which may be simply the end of a single phase of the system's life cycle, or may represent the final step in the overall system engineering design. In any case, the design process is carried out over and over until the program is terminated or the system reaches the end of its life cycle where, if a need still exists, the entire process may be started again.

System Life Cycle Phases

Figure 9-2 shows the different phases of the life cycle for a typical space system development. Each phase of the system life cycle has a definite *input* which, when subjected to the steps of the systems design process, leads to the *outputs* for the phase and ultimately for the period as well. Each phase is associated with the technical and programmatic (cost, risk, and schedule considerations) content and level of detail expected at that stage of mission development. As seen in the figure, a number of formal, documented reviews are used to ensure that the development is proceeding as expected and that all factors are such that the development can proceed to the next level. Considering the expense associated with such missions, and the usual difficulty of retrieving the satellite once placed into orbit, these factors include budgetary and political as well as quality assurance and technical considerations. The following sections will describe the level of detail expected, usual activities, and the reviews and/or decision gates associated with each phase in the unfolding development of a typical satellite mission.

Conceptual Study Phase (Pre-Phase A). This initial conceptual study phase is usually initiated by the intended users and mission advocates with the assistance of a spacecraft development organization (i.e., a NASA center, military organization, or spacecraft company). The phase begins by formulating a clear statement of the mission needs and objectives. Associated with the mission needs statement is a justification for the mission describing the scientific benefits that will be achieved, military goals accomplished, or projected profit to be made. Using descriptions of the

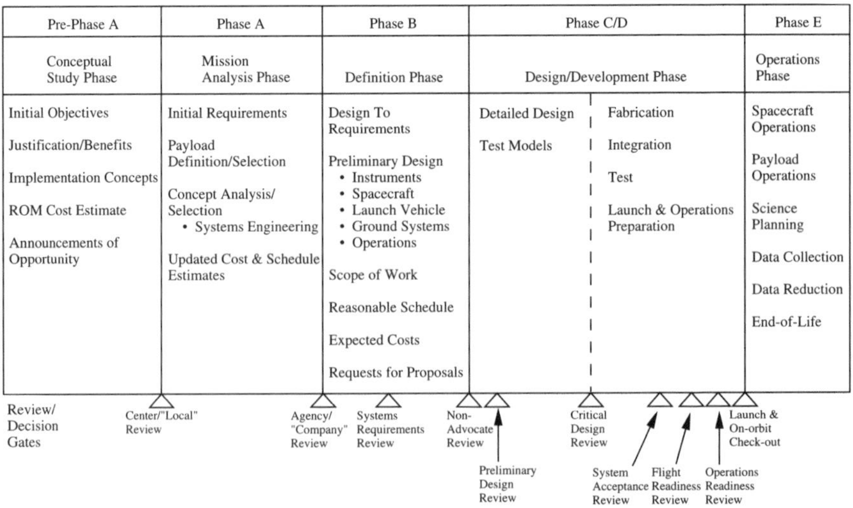

Pre-Phase A	Phase A	Phase B	Phase C/D		Phase E
Conceptual Study Phase	Mission Analysis Phase	Definition Phase	Design/Development Phase		Operations Phase
Initial Objectives	Initial Requirements	Design To Requirements	Detailed Design	Fabrication	Spacecraft Operations
Justification/Benefits	Payload Definition/Selection	Preliminary Design	Test Models	Integration	Payload Operations
Implementation Concepts	Concept Analysis/ Selection	• Instruments		Test	Science Planning
ROM Cost Estimate	• Systems Engineering	• Spacecraft • Launch Vehicle		Launch & Operations	Data Collection
Announcements of Opportunity	Updated Cost & Schedule Estimates	• Ground Systems • Operations		Preparation	Data Reduction
		Scope of Work			End-of-Life
		Reasonable Schedule			
		Expected Costs			
		Requests for Proposals			

Review/Decision Gates: Center/"Local" Review △ | Agency/"Company" Review △ Systems Requirements Review △ | Non-Advocate Review △ | Critical Design Review △ | Launch & On-orbit Check-out △△△

Preliminary Design Review | System Acceptance Review   Flight Readiness Review   Operations Readiness Review

Figure 9-2. System life cycle. The design of a spacecraft passes through many phases of evolution and review before becoming reality.

mission and payload, a number of different concepts may be formulated that can achieve the mission objectives, and these concepts will be expanded to determine the feasibility of conducting the mission. This concept development will include ground system and operations as well as launch vehicle and spacecraft descriptions, but the level of detail is usually in the form of estimates based on contemporary systems and technologies. After the concept is formulated, a *rough order of magnitude* (ROM) cost for the mission is attained using heritage cost models and known costs for subsystems and services. Exceptions to this are unique, one-of-a-kind designs or pioneering first efforts.

The results of this phase are reviewed at a "local" level—within a technical center, military or industrial laboratory, university, or similar organization—to determine if the concept warrants further consideration outside the group. For example, within NASA, if the concept looks promising, an *announcement of opportunity* (AO) may be issued which is the first step in defining and selecting the instruments/experiments that will fly on the mission if it is fully developed and launched. If concept study results prove to be promising, the evolving program moves on to the next step in Phase A, a more detailed mission evaluation.

Mission Analysis Phase (Phase A). The purpose of the mission analysis phase is to translate the broad mission concepts and objectives into a feasible preliminary system design. This is a refinement and expansion of the conceptual study phase made with the intention of providing a concise, clear overview of the proposed system. Implementation planning is done in cooperation with discipline engineering groups (attitude control, power, thermal, and other spacecraft subsystems areas defined in the preceding chapter), payload specialists (e.g., remote sensors and communications transponders), and ground and launch services organizations. Financial analysts utilize parametric cost models to derive top-level resource estimates for the overall mission and its component elements.

Key activities during this phase are system and subsystem trade studies, analyses of performance requirements, identification of advanced technology/long lead items, risk assessments, and end-to-end system life-cycle costs as a trade parameter. In addition, considerable attention is given to schedule options and selection and evaluation of operational concepts.

Specific tasks to support this phase may include technical feasibility and risk assessments associated with cost estimate modeling. Special attention is given to interface identification, definition, and analysis including spacecraft/ground data communications interfaces, satellite tracking system interfaces, and payload/spacecraft bus interfaces. The definition and evaluation of alternative strategies for the execution and management of the mission is also an important consideration.

The typical satellite Phase A study has a duration of approximately nine months to two years. Study costs are usually one to two percent of total mission costs.

Definition Phase (Phase B). The purpose of this phase is the refinement of the mission and system architecture and the system design created during the mission analysis phase. It is designed to convert the conceptual system design, which may still include some open trade-off and development options, into a final design and baseline for entering the execution phase. Functional, operational, and performance requirements are refined. Established interface requirements and specifications are allocated down to the subsystem or major component level. Firm cost estimates and schedules are prepared to assure a smooth transition to the subsequent execution phase C/D. Once a detailed baseline configuration that satisfies all the mission and program requirements has been established, the definition phase design, cost, and schedule are submitted to the program sponsor for approval.

Key activities during this phase are revalidation of mission requirements and system operational concepts. This phase also marks the conversion of mission requirements to much more specific project requirements. Final risk assessments are made along with more narrowly drawn system and subsystem studies and trades. Schedules and life-cycle costs are finalized to support the budget acquisition process.

This phase covers the full range of technical, management, resource, facility, and procurement assumptions. Phase B results in the full documentation of the entire proposed flight and ground support system. Alternate designs are analyzed in detail to allow the choice of one optimum flight and ground system approach considering technical performance, cost, risks, and schedules. The baseline design is totally defined as well as all major interfaces and system, subsystem, and component specifications.

Specific tasks conducted during this phase include telemetry assessments, analysis of payload hardware, launch vehicle interface analyses, and in-depth evaluation and categorization of the expected space environment. Components and subsystems are evaluated, as are production plans, facility requirements, and contingency workaround or offset plans.

This phase has a duration of from one to two years. The costs associated with this study are usually four to eight percent of total mission costs.

Design/Development Phase (Phase C/D). In many instances, these phases are combined into a single phase, since by now the commitment to build the system may already have been made. The purpose of the design/development phase is the implementation of the Phase B Plan. The final design, production development, and fabrication of hardware and software needed to translate the mission into a reality begins. The phase often includes launch, deployment, and postlaunch on-station system checkouts and follow-on operations. Depending on the scope of the deployed system (i.e., multiple satellite weather or navigation system developments versus one-of-a-kind single satellite programs) this phase can have an extremely variable duration, lasting from two years to more than a decade. The costs associated with this phase can also vary considerably—from tens of millions of dollars for SMALLSAT or LIGHTSAT missions to billions of dollars for defense or planetary exploration satellite programs.

Key activity categories during this phase are linked to the scheduled sequence of events, with each item listed below applicable to all the mission elements (space and ground systems and subsystems; payload elements including spacecraft/payload/ground system interfaces and links):

1. Design and development.
2. Fabrication.
3. Assembly.
4. Integration.
5. Test and evaluation (ambient and space environment exposures).
6. Transfer to the launch site followed by mating and prelaunch checkout.
7. Launch and launch sequence monitoring.
8. On-orbit checkout.
9. Postlaunch operations, which can include retrieval and on-orbit change or repair.

Perhaps the most important milestones in the Phase C/D process, shown in Figure 9-2, are the *preliminary design review* and the *critical design review* (PDR and CDR). These signify the release of the design for final production review (in the PDR), and the essential freeze of the design for fabrication and assembly (in the CDR).

Specific activities needed to support the sequence of events listed above include:

- Total quality management for reduction of mission problems (e.g., performance assurance, safety, reliability, maintainability, and quality assurance analyses and assessments).
- Assessment of specification packages to conform to all applicable standards.
- Evaluation of technical, progress, and feasibility reports.
- Liaison/coordination services to support project activities.
- Interface documentation and interface engineering.
- Mission plan.
- Operations and training manuals.
- Evaluation of configuration changes and changes to proposed designs or procedures.
- Monitoring fabrication of all hardware.
- Monitoring telemetry and command signal margins.
- Evaluation of logistical support for flight and ground support equipment.
- Review of functional test plans and procedures.
- Test and evaluation of the spacecraft, payload, and ground system.
- Flight and ground software verification and validation.
- Oversight of launch and operations support.
- Control of costs, schedules, and technical requirements.
- Plans for monitoring and controlling the allocations and error budget associated with power, thermal, attitude, timing, and mass properties, including launch margins.
- Readiness of systems for launch.
- Monitoring flight operations.
- Spacecraft/ground system command and data handling coordinated with the mission plan.
- Assessment of spacecraft/payload problems.

The outputs for this phase comprise all aspects of the space and ground segments of the mission and their operation.

Prior to proceeding to the operational phase, the launch must have been successful, the spacecraft should be deployed and in the correct orbit, the spacecraft and ground systems communications should be in normal working order, and the spacecraft should be operating successfully in concert with the tracking stations, telemetry and command, and data collection facilities.

Operations Phase (Phase E). The purpose of the operations phase is to safely and productively conduct the routine operation of the satellite, including data acquisition, processing, and distribution, until or unless a problem develops or the mission is terminated. The conduct of this phase varies considerably depending on whether the mission is a military, civil, scientific, or commercial endeavor.

Key activities during this phase are command and control of the spacecraft and instruments and data transmission, capture, processing, archiving, and distribution. Special mission operations may be authorized and conducted. Orbital changes may be required. Spacecraft or instrument components may be retrieved, repaired, or changed out in orbit. During this phase, routine items include:

- Payload performance reviews with users.
- Flight assessments of hardware, software, orbit behavior, and mission objectives.
- Monitoring the acquisition, processing, dissemination, and archiving of data.
- Compilation and production of a "lessons learned" document for the mission.

SPACECRAFT SYSTEMS

The following is a list of factors that must be considered in the design of systems which are to perform in the environment of space. The list is in no way complete or exhaustive but represents a minimum of topics which must be considered in such systems. These factors were presented in detail in Chapter 8 along with their associated design considerations.

A. Space Component
　1. Payload (mission-specific) systems
　2. Support systems
　　a. power systems
　　b. thermal control
　　c. propulsion
　　d. attitude reference and control
　　e. data
　　f. spacecraft monitoring and control
　　g. structure

B. Ground Support Components
1. Spacecraft design and construction
2. Launch sites
   a. launch vehicles
   b. upper stages
3. Ground Sites
C. User Components

REFERENCES/ADDITIONAL READING

Wertz, J. and Larson, W., *Space Mission Analysis and Design.* Boston: Kluwer Academic Publishers, 1991.

Griffin, M. and French, J., *Space Vehicle Design.* Washington, D.C.: American Institute of Aeronautics and Astronautics, 1991.

The Flight Projects Directorate Project Management Handbook, NASA/ Goddard Space Flight Center, 1994.

EXERCISES

1. Choose one of the common uses of systems in space (communications, navigation, remote sensing) and complete the following:
   a. Define the need for such a system and justify its performance from a space-based solution versus a ground-based system.
   b. Define the top-level characteristics (requirements) for the system. This is a listing of what must be accomplished to meet the stated need, but does not assign functions to any particular segment of the system (space segment, ground segment, etc.).
   c. Create a value model using these characteristics. Assign and provide rationale for weighting factors for each characteristic.
   d. For each of the space system components listed at the end of this chapter, take a first cut on assignment of functions and component characteristics. This will involve defining the basic mission scenario, including launch site and vehicle, ground systems and operation, as well as defining the types of spacecraft subsystems appropriate for the mission. Assumptions may be made for the payload (size, mass, power, data, etc.).
   e. The next step would be to perform an initial sizing of the system and subsystems. Good luck!

Manned Spaceflight Summary

The following table summarizes United States and Soviet manned space launches up to the first launch of the Space Shuttle.

Designation	Launch	Recovery	Astronauts	Notes
Vostok 1	12 Apr 61	12 Apr 61	Gagarin	1 orbit, 1.8 hr.
Mercury-Redstone 3	5 May 61	5 May 61	Shepard	Suborbital, 15 min.
Mercury-Redstone 4	21 Jul 61	21 Jul 61	Grissom	2nd U.S. suborbital
Vostok 2	6 Aug 61	7 Aug 61	Titov	17 orbits, 25 hrs.
Mercury-Atlas 6	20 Feb 62	20 Feb 62	Glenn	3 orbits, 4.9 hrs.
Mercury-Atlas 7	24 May 62	24 May 62	Carpenter	3 orbits, 4.9 hrs.
Vostok 3	11 Aug 62	15 Aug 62	Nikolayev	64 orbits, 94.4 hrs.
Vostok 4	12 Aug 62	15 Aug 62	Popovich	Came within 5 km of *Vostok 3.*
Mercury-Atlas 8	3 Oct 62	3 Oct 62	Schirra	6 orbits, 9.2 hrs.
Mercury-Atlas 9	15 May 63	16 May 63	Cooper	22 orbits, 34.3 hrs.
Vostok 5	14 Jun 63	19 Jun 63	Bykovsky	81 orbits, 119.1 hrs.
Vostok 6	16 Jun 63	19 Jun 63	Tereshkova	Came within 5 km of *Vostok 5.* First woman in space.
Voskhod 1	12 Oct 64	13 Oct 64	Komarov Feokistov Yegorov	16 orbits. First multiple crew.
Voskhod 2	18 Mar 65	19 Mar 65	Leonov Belyayev	First EVA (Leonov, 20 min.)

(table continued on next page)

211

Designation	Launch	Recovery	Astronauts	Notes
Gemini 3	23 Mar 65	23 Mar 65	Grissom Young	Orbital maneuvers.
Gemini 4	3 Jun 65	7 Jun 65	White McDivitt	21 min. EVA (White).
Gemini 5	21 Aug 65	29 Aug 65	Cooper Conrad	128 orbits, 190.9 hrs.
Gemini 7	4 Dec 65	18 Dec 65	Borman Lovell	220 orbits, 330.6 hrs.
Gemini 6	15 Dec 65	16 Dec 65	Schirra Stafford	Rendezvous (RVZ) within 1 ft of *Gemini 7*.
Gemini 8	16 Mar 66	16 Mar 66	Armstrong Scott	Docked with Agena but suffered control failure.
Gemini 9	3 Jun 66	6 Jun 66	Stafford Cernan	Fairing prevented docking. Conducted RVZ and EVA.
Gemini 10	18 Jul 66	21 Jul 66	Young Collins	Docked and used Agena to raise orbit altitude.
Gemini 11	12 Sep 66	15 Sep 66	Conrad Gordon	Docked with Agena within 1st orbit.
Gemini 12	11 Nov 66	15 Nov 66	Lovell Aldrin	63 orbits, 94.6 hrs. Multiple EVAs.
Apollo 1			Grissom White Chaffee	Astronauts killed in fire on launch pad during tests 27 Jan 67.
Soyuz 1	23 Apr 67	24 Apr 67	Komarov	Cosmonaut killed during recovery.
Apollo 7	11 Oct 68	22 Oct 68	Schirra Cunningham Eisele	163 orbits, 260.2 hrs.
Soyuz 3	28 Oct 68	30 Oct 68	Beregoyov	RVZ to 198 m (650 ft) of unmanned *Soyuz 2*.
Apollo 8	21 Dec 68	27 Dec 68	Borman Lovell Anders	10 lunar orbits.
Soyuz 4	14 Jan 69	17 Jan 69	Shatalov	Docked with *Soyuz 5*.
Soyuz 5	15 Jan 69	18 Jan 69	Volynov Khrunov Yeliseyev	EVA transferred 2 cosmonauts to *Soyuz 4* for landing. Volynov recovered alone.
Apollo 9	3 Mar 69	13 Mar 69	McDivitt Scott Schweickart	Tested lunar module in earth orbit.
Apollo 10	18 May 69	26 May 69	Young Stafford Cernan	Lunar landing rehearsal piloted LM to 15 km (9.3 mi) of lunar surface.

Designation	Launch	Recovery	Astronauts	Notes
Apollo 11	16 Jul 69	24 Jul 69	Armstrong Aldrin Collins	Sea of Tranquility, 21.6 hrs. on lunar surface. 2.5 hr. lunar EVA.
Soyuz 6	11 Oct 69	16 Oct 69	Shonin Kubasov	3 simultaneous flights (Soyuz 6, 7, and 8). Conducted welding tests.
Soyuz 7	12 Oct 69	17 Oct 69	Filipchenko Volkov Gorbatko	Navigation/photo experiments.
Soyuz 8	13 Oct 69	18 Oct 69	Shatalov Yeliseyev	RVZ with *Soyuz 7*. Comm./photo experiments.
Apollo 12	14 Nov 69	24 Nov 69	Conrad Bean Gordon	Ocean of Storms, 31.6 hrs. on lunar surface. 2 EVAs. Recovered Surveyor parts.
Apollo 13	11 Apr 70	17 Apr 70	Lovell Swigert Haise	O_2 tank rupture crippled service module. Crew in LM circled moon to return.
Soyuz 9	2 Jun 70	19 Jun 70	Nikovayev Sevastianov	17 days, 16 hrs., 59 min.
Apollo 14	31 Jan 71	9 Feb 71	Shepard Mitchell Roosa	Fra Mauro area, 2 EVAs, 9 hrs.
Salyut 1	19 Apr 71	11 Oct 71		Prototype space station.
Soyuz 10	23 Apr 71	24 Apr 71	Shatalov Yeliseyev Rukavishnikov	Docked with *Salyut 1* but did not enter.
Soyuz 11	6 Jun 71	29 Jun 71	Dobrovolsky Volkov Patsayev	Manned *Salyut 1* (22 days). Crew perished during recovery.
Apollo 15	26 Jul 71	7 Aug 71	Scott Irwin Worden	Hadley-Apennines area. Used lunar roving vehicle. Left satellite in lunar orbit.
Apollo 16	16 Apr 72	27 Apr 72	Young Duke Mattingly	Descartes area. EVA on return from moon to retrieve film cannister.
Apollo 17	7 Dec 72	19 Dec 72	Cernan Schmitt Evans	Taurus-Littrow area. Last lunar mission. Schmitt first scientist/astronaut.
Salyut 2	3 Apr 73	28 May 73		Space station (failure).
Skylab 1	14 May 73	11 Jul 79		U.S. space station.

(table continued on next page)

Designation	Launch	Recovery	Astronauts	Notes
Skylab 2	25 May 73	22 Jun 73	Conrad Kerwin Weitz	28 days. Conducted unscheduled EVAs to repair *Skylab* damage during launch.
Skylab 3	28 Jul 73	25 Sep 73	Bean Garriott Lousma	59 days. More *Skylab* repairs. Extensive earth and sun observations.
Soyuz 12	27 Sep 73	29 Sep 73	Lazarev Makarov	Test of Soyuz redesign.
Skylab 4	16 Nov 73	8 Feb 74	Carr Gibson Pogue	84 days. Longest U.S. space mission to date.
Soyuz 13	18 Dec 73	26 Dec 73	Klimuk Lebedev	Astronomical experiments.
Salyut 3	24 Jun 74	24 Jan 75		Space station.
Soyuz 14	3 Jul 74	19 Jul 74	Popovich Artyukhin	Manned *Salyut 3*.
Soyuz 15	26 Aug 74	28 Aug 74	Sarafanov Demin	Rendezvous but did not man *Salyut 3*.
Soyuz 16	2 Dec 74	8 Dec 74	Filipchenko Rukavishnikov	Test of ASTP spacecraft.
Salyut 4	26 Dec 74	2 Feb 77		Space station.
Soyuz 17	10 Jan 75	9 Feb 75	Gubarev Grechko	Manned *Salyut 4*.
Soyuz 18-A?	5 Apr 75	5 Apr 75	Lazarev Makarov	Booster failure aborted mission after launch.
Soyuz 18	24 May 75	26 Jul 75	Klimuk Sevastyanov	Manned *Salyut 4*.
Soyuz 19	15 Jul 75	21 Jul 75	Leonov Kubasov	American/Soviet Test Program (ASTP).
Apollo 18	15 Jul 75	24 Jul 75	Stafford Brand Slayton	ASTP.
Soyuz 20	17 Nov 75	16 Feb 76		Unmanned Soyuz docked to *Salyut 4*.
Salyut 5	22 Jun 76	8 Aug 77		Space station.
Soyuz 21	6 Jul 76	24 Aug 76	Volynov Zholobov	Manned *Salyut 5*.
Soyuz 22	15 Sep 76	23 Sep 76	Bykovsky Aksenov	Independent flight.
Soyuz 23	14 Oct 76	16 Oct 76	Zudov Rozhdestensky	Failed to dock with *Salyut 5*.

Designation	Launch	Recovery	Astronauts	Notes
Soyuz 24	7 Feb 77	25 Feb 77	Gorbatko Glazkov	Manned *Salyut 5*.
Salyut 6	29 Sep 77			Space station.
Soyuz 25	9 Oct 77	11 Oct 77	Kovalenok Ryumin	Failed to dock with *Salyut 6*.
Soyuz 26	10 Dec 77	16 Jan 78	Romanenko Grechko	Manned *Salyut 6* for 96 days. Returned in *Soyuz 27* s/c.
Soyuz 27	10 Jan 78	16 Feb 78	Dzhanibekov Makarov	Visited *Salyut 6* crew 5 days. Returned in *Soyuz 26* s/c.
Progress 1	20 Jan 78	8 Feb 78		Progress craft are unmanned supply ships to Soviet space stations.
Soyuz 28	2 Mar 78	10 Mar 78	Gubarev Remek (Czech)	Manned *Salyut 6*.
Soyuz 29	15 Jun 78	3 Sep 78	Kovalenok Ivanchenkov	Manned *Salyut 6* 139 days. Returned in *Soyuz 31* s/c.
Soyuz 30	27 Jun 78	5 Jul 78	Klimuk Hermanszewski (Polish)	Visited *Salyut 6* crew.
Soyuz 31	26 Aug 78	2 Nov 78	Bykovskiy Jaehn (E. German)	Returned in *Soyuz 29* s/c.
Soyuz 32	25 Feb 79	13 Jun 79	Lyakhov Ryumin	Manned *Salyut 6* 175 days. Returned in *Soyuz 34* s/c.
Soyuz 33	10 Apr 79	12 Apr 79	Rukavishnikov Ivanov (Bulgarian)	Failed to dock with *Salyut 6*.
Soyuz 34	6 Jun 79	19 Aug 79		Launched unmanned, returned *Salyut 6* crew.
Soyuz T-1	16 Dec 79	25 Mar 80		Launched unmanned, docked with *Salyut 6*.
Soyuz 35	9 Apr 80	3 Jun 80	Popov Ryumin	Manned *Salyut 6* 185 days. Returned in *Soyuz 37* s/c.
Soyuz 36	26 May 80	30 Jul 80	Kubasov Farkash (Hungarian)	Returned in *Soyuz 35* s/c.
Soyuz T-2	5 Jun 80	24 Jun 80	Malyshev Aksenov	Modified Soyuz spacecraft. Docked with *Salyut 6*.

(table continued on next page)

Designation	Launch	Recovery	Astronauts	Notes
Soyuz 37	23 Jul 80	11 Oct 80	Gorbatko Pham (Vietnamese)	Returned in *Soyuz 36* s/c.
Soyuz 38	18 Sep 80	26 Sep 80	Romanenko Tamayo (Cuban)	Manned *Salyut 6*.
Soyuz T-3	27 Nov 80	10 Dec 80	Kizin Makarov Strekalov	3-man "civilian" mission to *Salyut 6*.
Soyuz T-4	12 Mar 81	26 May 81	Kovalyenok Savinykh	Manned *Salyut 6*.
Soyuz 39	22 Mar 81		Dzhanibekov Gurragcha (Mongolian)	Manned *Salyut 6*.
STS 1	12 Apr 81	14 Apr 81	Young Crippen	First Space Shuttle flight.

United States and Foreign Spacecraft Inventories and Spacecraft Descriptions*

U. S. SPACECRAFT

Spacecraft	1st Launch	Launch Vehicle	Mission
AEROS	16 Dec 1972	Scout	Upper atmosphere research.
AFSATCOM	19 May 1979	Atlas Centaur	Air Force Communications (part of FLTSATCOM).
Air Density Explorer	19 Dec 1963	Scout	Explorers 19, 24, 25, 39.
AMPTE (Active Magnetosphere Particle Trace Explorer)	16 Aug 1984	Delta 3924	Magnetospheric research.
ANNA (Army, Navy, NASA, Air Force) satellite	31 Oct 1962	Thor Able	Geodetic research (see Secor).
Apollo (Manned)	11 Oct 1968	Saturn	Lunar exploration, Skylab, Soyuz.
AEM-A (Applications Explorer Mission)	26 Apr 1978	Scout	Visual/IR earth obs. (HCMM).
AEM-B	18 Feb 1979	Scout	Solar radiation study (SAGE).

Not intended to be a comprehensive listing.

(table continued on next page)

Spacecraft	1st Launch	Launch Vehicle	Mission
ASC (American Satellite Co.)	27 Aug 1985	Shuttle	Private commercial telecommunication.
ATS (Applications Technology Satellites) (ATS 1 thru 6)	7 Dec 1966	Atlas Agena	Geosynchronous experimental satellites.
Atmosphere Explorer	29 Apr 1965	Scout	Explorers 27, 32, 51, 54, 55.
Beacon	23 Oct 1968	Jupiter C (Juno)	Atmospheric research balloons.
Beacon Explorers	10 Oct 1964	Delta, Scout	Explorers S-66, 22, 27 (geodetic).
Big Bird (Broad Coverage Photo Reconnaissance Satellites)	Jun 1971	Titan 3D/ Agena	Photo/radio reconnaissance.
Biosatellite	15 Feb 1967	Delta	Biological research.
BMEWS (Ballistic Missile Early Warning System)	9 May 1963	Atlas Agena	Missile launch detection (MIDAS).
Calsphere	6 Oct 1968	Thor	Early experimental ASAT.
Cameo (Chemically Active Material into Orbit)	24 Oct 1978	Delta 2910	Rel. by Nimbus 7 to study aurora.
COBE (Cosmic Background Explorer)	1989	Shuttle	Big Bang radiation measurement.
COMSAT (Communications Satellite Corporation)			A company, not a satellite.
Comstar	13 May 1976	Atlas Centaur	COMSAT telephone communications satellite.
Courier	4 Oct 1960	Atlas B	Post-SCORE Army communications satellite.
CRL (Cambridge Research Laboratory)	28 Jun 1963	Scout	Geophysics research.
CRRS (Combined Radiation and Release Satellite)	1985	Shuttle	Radiation effects measurement.
Discoverer	28 Feb 1959	Thor Agena	Early reconnaisance satellites.
DMSP (Defense Meteorological Satellite Program)	Oct 1971	Atlas E	Military meteorological satellites.

Spacecraft	1st Launch	Launch Vehicle	Mission
DSCS (Defense Satellite Communications System)	16 Jun 1966	Titan 3C	Global strategic military communications.
DSP (Defense Support Program)	5 May 1971	Titan 3C	Ballistic missile launch detection.
Dynamics Explorer	3 Aug 1981	Delta	Near-earth space environment.
Early Bird	6 Apr 1965	Delta	Commercial communications (Intelsat 1).
Echo	12 Aug 1960	Thor Delta	Passive communications balloon.
ELINT (Electronic Inteligence) satellites	Mar 1962	Various (Hitchikers)	Electronic information gathering.
ERBS (Earth Radiation Budget Satellite)	5 Oct 1984	Shuttle	Solar radiation measurement.
ERTS (Earth Resources Technology Satellite)	23 Jul 1972	Delta	Visual/IR earth observation (Landsat).
ESSA (U.S. Environmental Science Services Administration) satellites	3 Feb 1966	Delta	TOS (TIROS Operational Satellites) (TIROS 1960–65, TOS/ESSA 1966–69, ITOS/NOAA 1970–84, TIROS-N 1978).
EUVE (Extreme Ultraviolet Explorer)	1988	Shuttle	Ultraviolet telescope.
Explorer	31 Jan 1958	Jupiter C	1st American satellite in orbit; Series of 61 scientific research satellites from 1958 thru 1975.
Ferret (Code 711)	pre-1973	Thor Agena	Electronic reconnaissance precursor to Rhyolite.
FLTSATCOM (U.S. Navy Fleet Satellite Communications)	9 Feb 1973	Atlas Centaur	Global UHF military communications.
Galaxy	28 Jun 1983	Delta	Domestic T.V.
Galileo	1989	Shuttle	Jupiter orbiter.
Gemini (manned)	23 Mar 1965	Titan 2	Precursor to Apollo.
Geodetic Explorer	6 Nov 1965	Delta	Explorers 29, 36 (GEOS A,B).
GEOS (Geodynamic Experimental Ocean Satellite)	9 Apr 1975	Delta	Measurement of earth shape.

(table continued on next page)

Spacecraft	1st Launch	Launch Vehicle	Mission
Geosat	12 Mar 1985	Atlas E	Measurement of earth shape.
GOES (Geostationary Operational Environmental Satellite)	16 Oct 1975	Delta 2914	Global meteorological information.
GPS (Global Positioning System)	22 Feb 1978	Atlas E/F	3-D position and speed fix (NAVSTAR).
GRO (Gamma Ray Observatory)	1988	Shuttle	Gamma ray telescope.
G-STAR	May 1985	Ariane 3	U.S. commercial domestic communications.
HCMM (Heat Capacity Mapping Mission)	26 Apr 1978	Scout D	Visual/IR earth observations.
HEAO (High Energy Astronomy Observatory)	12 Sep 1977	Atlas Centaur	Celestial X-ray, Gamma ray map.
Helios	10 Dec 1974	Titan 3E	Close solar observation.
Hilat	27 Jun 1983	Scout	Ionospheric research.
HS 376	11 Nov 1982	Shuttle (or others)	Hughes commercial communications satellite. (Galaxy, SBS, Telstar, Westar).
HST (Hubble Space Telescope)	1989	Shuttle	Orbiting optical telescope.
IMEWS (Integrated Missile Early Warning Satellite)	5 May 1971	Titan 3C	Geostationary infrared detection.
Intelsat (International Telecommunications Satellite Consortium)			In cooperation with COMSAT
Intelsat (International Telecommunications Satellite Consortium) satellites	6 Apr 1965	Delta, Atlas Centaur	Series of communications satellites (Blocks 1-6).
IUE (International Ultraviolet Explorer)	26 Jan 1978	Delta	Celestial UV observation.
IMP (Interplanetary Monitoring Platform)	26 Nov 1963	Delta	Measure interplanetary environment (Explorers 18, 21, 28, 33–35, 41, 43, 47, 50).
Ionds	22 Feb 1978		Nuclear detection satellite.
Ionosphere Explorer	25 Aug 1964	Scout	Upper atmospheric observation (Explorer 20).

Spacecraft	1st Launch	Launch Vehicle	Mission
IRAS (Infrared Astronomical Satellite)	25 Jan 1983	Delta	Celestial infrared observation.
ISEE (International Sun-Earth Explorer)	22 Oct 1977	Delta	Earth-sun space environment.
ISIS (International Satellites for Ionospheric Studies)	30 Jan 1969	Delta	Ionospheric observation.
ITOS (Improved TIROS Operational System)	23 Jan 1970	Delta N	Weather satellites (TIROS 1960–65, TOS/ESSA 1966–69, ITOS/NOAA 1970–84, TIROS-N 1978).
KH-9/11/12 (Key Hole)	29 Jul 1966	Titan 3D/ Agena	Photo reconaissance satellites.
LAGEOS (Laser Geodynamics Satellite)	4 May 1976	Delta 2914	Passive laser reflector for ranging.
Landsat (ERTS Earth Resources Technology Satellite)	23 Jul 1972	Delta	Visual and IR earth observation.
Leasat	1 Sep 1984	Shuttle	Leased FLTSATCOM capability.
LDEF (Long Duration Exposure Facility)	8 Apr 1984	Shuttle	Space exposure data.
LOFTI (Low Frequency Trans- Ionospheric Satellite)	21 Feb 1961	Thor (Piggyback)	Navigation satellite research.
Lunar Orbiter	10 Aug 1966	Atlas Agena	Lunar topographical data.
Lunar Roving Vehicle			Moon Buggy.
Magsat (Magnetic Field Satellite)	30 Oct 1979	Scout	Near-earth magnetic field data.
Mariner	22 Jul 1962	Atlas Agena/ Centaur	Planetary exploration.
Marisat	19 Feb 1976	Delta	Maritime communications.
Mercury	5 May 1961	Redstone/ Atlas	1st American manned missions.
Meteoroid Satellites	30 Jun 1961	Scout	Explorers S-55, 13, 16, 23, 46.
MIDAS (Missile Alarm Defense System)	26 Feb 1960	Atlas Agena	IMEWS precursor (BMEWS).
MILSTAR	1988	Shuttle	Military communications.
NATO (NATOSAT)	20 Mar 1970	Delta	NATO geostationary communications satellites.

(table continued on next page)

Spacecraft	1st Launch	Launch Vehicle	Mission
Navsat (Naval Navigation Satellite)			Operational name of Transit system.
NAVSTAR (Navigation System Using Time and Ranging)/GPS	22 Feb 1978	Atlas E/F	3-D position and speed fixing.
NERVA (Nuclear Engine for Rocket Vehicle Application)			Nuclear-powered heavy lift vehicle (see RFD)
Nimbus	28 Aug 1964	Thor Agena/ Delta	Weather satellites.
NOAA (National Oceanic and Atmospheric Administration) satellites	27 Jun 1979	Atlas	ITOS meteorological follow-on (TIROS 1960–65, TOS/ESSA 1966–69, ITOS/NOAA 1970–84, TIROS-N 1978).
NOSS	30 Apr 1976		Naval ocean surveillance system.
NOVA	15 May 1981	Scout	Improved Transit navigational satellite (TIP).
NTS (Navigation Technology Satellites)	14 Jul 1974	Atlas F	Timation/Transit improvement.
OAO (Orbiting Astronomical Observatory)	8 Apr 1966	Atlas Agena/ Centaur	Observation of interstellar space.
OFO (Orbiting Frog Otolith)	9 Nov 1970	Scout	Biological experiments on frogs in space.
OGO (Orbiting Geophysical Observatory)/POGO	6 Sep 1964	Atlas Agena	Study of geophysical and solar phenomena.
OSCAR (Orbiting Satellite-Carrying Amateur Radio)	23 Jan 1970	Piggyback	Ham radio operator broadcasts.
OSO (Orbiting Solar Observatory)	7 Mar 1962	Delta	Sun and solar physics observation.
PAGEOS (Passive GEOS)	24 Jun 1966	Thor Agena	Sunlight reflector for positioning.
Pegasus	16 Feb 1965	Saturn 1/ Apollo	Micrometeorite measurements.
Pioneer	11 Oct 1958	Thor, Juno, Atlas, Delta, Atlas Centaur	Lunar, solar, and planetary (Jupiter, Mars, and Venus) probes.
Probe A/B	19 Oct 1961	Scout	Electron density measurement.
Project 78	24 Feb 1979	Atlas	Earth/sun gamma-ray and solar dyn.

Spacecraft	1st Launch	Launch Vehicle	Mission
RMS (Radiation Meteoroid Satellite)	9 Nov 1970	Scout	Meteoroid direction, speed, and flux.
Radio Astronomy Explorer			Explorer 38, 49.
RAM Re-entry Experiments	19 Oct 1967	Scout	Reentry effects on communication.
Ranger	23 Feb 1961	Atlas Agena	Lunar pictures until impact.
RCA Satcom	13 Dec 1975	Delta 3914	Commercial communications/TV.
Reentry	1 Mar 1962	Scout	Reentry effects on ablative material.
Relay	13 Dec 1962	Thor Delta	Low-altitude repeater comsat.
RFD (Re-entry Flight Demonstrator) (NERVA)	22 May 1963	Scout	Reentry effect on nuclear reactor.
Rhyolite	6 March 1963		Geosynchronous electromagnetic reconnaissance.
SAGE (Stratospheric Aerosol and Gas Experiment)	18 Feb 1979	Scout	Solar radiation in earth atmosphere.
SAMOS (Satellite and Missile Observation System)	11 Oct 1960	Atlas, Titan Agena	Polar orbiting TV reconnaissance.
Satcom			RCA Satcom.
SBS (Satellite Business System)	15 Nov 1980	Delta, Shuttle	Digital business telecommunication.
SCATHA (Spacecraft Charging at High Altitudes)	30 Jan 1979	Delta 2914	Geomagnetic field effects on s/c.
SCORE (Signal Communication by Orbiting Relay Equipment)	18 Dec 1958	Atlas B	Merry Christmas from President Ike.
Seasat	26 Jun 1978	Atlas F	Active/passive oceanographic.
Secor (Sequential Collation of Range spacecraft)	11 Jan 1964	Thor Agena	Army navigation/positioning.
SERT (Space Electric Rocket Test)	20 Jul 1964	Scout, Thor Agena	Ion engine experiments.
Skylab	14 May 1973	Saturn V	1st U. S. space station.
Small Astronomy Satellite			Explorer 53.

(table continued on next page)

Spacecraft	1st Launch	Launch Vehicle	Mission
Small Scientific Satellite			Explorer 45.
SME (Solar Mesosphere Explorer)	6 Oct 1981	Delta	Atmospheric research.
SMM (Solar Maximum Mission)	14 Feb 1980	Delta 3910	Solar flare, radiation, particle observation.
SMS (Synchronous Meteorological Satellite)	17 May 1974	Delta	First GOES satellite.
Solar Explorer			Explorer 30, 37.
SOLRAD			Explorer 44.
Spacenet	23 May 1984	Ariane	U.S. commercial domestic communications.
STC (Satellite Television Corp.)	1987	Ariane	U.S. television broadcasts.
STS (Space Shuttle)	12 Apr 1981	SRBs	Space Transportation System.
Squanto Terror	Mar 1964	Thor (Johnson Is.)	Once-operational nuclear ASAT.
Surveyor	30 May 1966	Atlas Centaur/Agena	Soft lunar lander.
SMS (Synchronous Meteorologica/ Satellites)	17 May 1974	Delta	Precursors of GOES.
Syncom	14 Feb 1963	Thor Delta	1st geosynchronous communications satellites.
Tacsat	9 Feb 1969	Titan 3C	Military tactical communications.
TDRS (Tracking and Data Relay Satellite)	4 Apr 1983	Shuttle/IUS	S/C tracking and data relay.
Teal Ruby	1987		IR aircraft detection system.
Telstar	10 Jul 1962	Thor Delta	1st commercial communications satellite.
Timation (Time Navigation)	May 1967	Scout	Precursor to GPS.
TIP (Transit Improved Program)/NOVA	2 Sep 1972	Scout	Navigational satellite with DISCOS/RTGs.
TIROS (Television Infrared Observation Satellite)	1 Apr 1960	Thor, Delta, Atlas	Infrared weather monitoring (TIROS 1960–65, TOS/ESSA 1966–69, ITOS/NOAA 1970–84, TIROS-N 1978).
TOPO	8 Apr 1970	Thor Agena	Army space-ground triangulation.

Spacecraft	1st Launch	Launch Vehicle	Mission
TOS (TIROS Operational System)	3 Feb 1966	Delta	ESSA meteorological satellites (TIROS 1960–65, TOS/ESSA 1966–69, ITOS/NOAA 1970–84, TIROS-N 1978).
TRAAC (Transit Research and Attitude Control) satellite	15 Nov 1961	Thor (Piggyback)	Gravity gradient research.
Transit	Sep 1959	Scout	Navigational satellite system.
UARS (Upper Atmosphere Research Satellite)	1989	Shuttle	Stratosphere, mesosphere monitor.
Vanguard	6 Dec 1963	Vanguard TV-3	Early Navy satellite program.
Vela	17 Oct 1963	Atlas Agena	Nuclear explosion detection.
Viking	20 Aug 1975	Titan Centaur	Soft landing on Mars.
Voyager	20 Aug 1977	Titan Centaur	Planetary exploration.
Westar	13 Apr 1974	Delta 2914	1st domestic geosynchronous communications satellites.
Whitecloud	30 Apr 1976	Atlas	Electronic ocean surveillance/ tracking.

NON-U. S. SATELLITES

Spacecraft	1st Launch	Launch Vehicle	Mission
Aeros	16 Dec 1972	Scout	W. German environment tests.
Alouette	29 Sep 1962	Thor Agena	Canadian ionospheric effects on communications
Anik-B	15 Dec 1978	Delta	Canadian telecommunications.
ANS (Astronomical Netherlands Satellite)	30 Aug 1974	Scout	Netherland X-ray studies.
Apple (Ariane Passenger Payload Experiment)	19 Jun 1981	Ariane	Indian communications experiments.
Arabsat	8 Feb 1985	Ariane, Shuttle	Arab League telecommunications.
Ariel	27 Mar 1964	Scout	U. K. radio and environment tests.
Aryabhata	19 Apr 1975	C-1	Indian high-altitude physics.
Astron	23 Mar 1983	D-1	Soviet X-ray/UV telescope.
Aussat	27 Aug 1985	Shuttle	Australian telecommunications.

(table continued on next page)

Spacecraft	1st Launch	Launch Vehicle	Mission
Ayame/ECSs (Experimental Communications Satellite)	6 Feb 1979	N-1	Japanese communications.
AZUR	8 Nov 1969	Scout	W. German particle measurements.
Bhaskara	7 Jun 1979	C-1 (Sov)	Indian Earth resources satellite.
Brasilsat	8 Feb 1985	Ariane	Brazilian telecommunications.
China-1	24 Apr 1970	Long March-1	Chinese space environment tests.
CORSA (Cosmic Radiation Satellite)	21 Feb 1979	M-3C-4	Japanese cosmic radiation experiment.
COS-B	9 Aug 1975	Delta	ESA gamma ray observations.
COSMOS-1076	12 Feb 1979		Soviet oceanographic satellite. (Most Soviet spacecraft are given the generic label COSMOS until typed).
Cosmos 1629/1700	21 Feb 1985		Soviet TDRS-type relay satellite.
Cosmos-Meteor	25 Jun 1966	A-2	Soviet weather satellite.
Cosmos-Tsikada	23 Nov 1967	C-1	Soviet Transit-type navigational satellite.
CTS (Communications Technology Satellite)/ Hermes	17 Jan 1976	Delta 2914	USA/Canada/ESA communications satellite.
DENPA/REXS	19 Aug 1972	M-4S-4	Japanese magnetic field experiment.
DIAL/WIKA	10 Mar 1970	Diamant B	W. German particle measurements.
Diapason/Diademe	17 Feb 1966	Diamant	French geodetic satellites.
ECS (European Communications Satellite)	16 June 1983	Ariane	European continental communications satellite.
EGP (Experimental Geodetic Payload)	8 Aug 1986	H-1	Japanese geodetic research.
Ekran	26 Oct 1976	D-1	Soviet TV broadcast.
Elektron	30 Jan 1964	A-1	Soviet radiation belt experiments.
EOLE FR-2	16 Aug 1971	Scout (Vandenberg)	French atmospheric data gathering.
ERS (European Remote Sensing) satellite	1989	Ariane	European remote sensing satellite.
ESRO (European Space Research Organization)	17 May 1968		ESA space physics experiments.

Spacecraft	1st Launch	Launch Vehicle	Mission
ETS/Kiku (Engineering Test Satellite)	23 Feb 1977	N-1	1st Japanese geostationary test satellite.
Exos	4 Feb 1978	M-3H-2	Japanese particle measurements.
EXOSAT (European X-ray Observatory Satellite)	26 May 1983	Delta 3914	ESA X-ray observations.
FR-1	6 Dec 1965	Scout	French ionosphere/magnetosphere.
GEOS	20 Apr 1977	Delta	ESA magnetospheric studies.
GLONASS (Global Navigation Satellite System)	12 Oct 1982	D-1	Soviet GPS-type navigation system.
GMS/Himawari (Geostationary Meteorological Satellite)	14 Jul 1977	Delta	Japanese meteorological satellite.
Gorizont	19 Dec 1978	D-1	Soviet geostationary communications/TV satellite.
HEOS (Highly Eccentric Orbit Satellite)	5 Dec 1968	Delta	ESA solar/magnetosphere experiment.
Hinotori/Astro	21 Feb 1981	M-3S-2	Japanese solar flare observations.
INSAT-B (Indian National Satellite)	30 Aug 1983	Shuttle	Indian meteorological/ communications satellite.
Intasat	15 Nov 1974	Delta	Spanish signal propagation experiment.
ISEE-2 (International Sun-Earth Explorer)	22 Oct 1977	Delta	ESA portion of ISEE experiment.
ISIS (International Satellites for Ionospheric Studies)	30 Jan 1969	Thor Delta	Alouette follow-on.
Italsat	1989	Ariane	Italian telecommunications.
JERS (Japanese Earth Resources Satellite)	1991	H-1	Japanese SAR remote sensor.
Marecs	20 Dec 1981	Ariane	ESA maritime communications.
Meteor-1	26 Mar 1969	A-2	Soviet weather satellite.
Meteor-2	11 Jul 1975	A-1	Soviet meteorological satellite.

(table continued on next page)

Spacecraft	1st Launch	Launch Vehicle	Mission
Meteor-Priroda	10 Jul 1981	A-1	Soviet earth resources satellite.
Meteosat	23 Nov 1977	Delta 2914	ESA geosynchronous meteorological satellite.
Molniya	23 Apr 1965	A-2	Soviet telecommunications.
Morelos	17 Jun 1985	Shuttle	Mexican telecommunications.
MOS (Marine Observations Satellite)	1987	N-2	Japanese Earth observation satellite.
Olympus (L-Sat)	1988	Ariane 3	ESA advanced telecommunications experiment.
OREOL/AUREOLE	27 Dec 1971	C-1	Soviet/French solar wind experiment.
OTS (Orbital Test Satellite)	12 May 1978	Delta 3914	ESA experimental communications satellite.
Palapa	8 Jul 1976 18 Jun 1983	Delta Shuttle	Indonesian telecommunications.
Peole	12 Dec 1970	Diamant B	French geodetic/transmission tests.
Pollux/Castor	17 May 1975	Diamant B-P4	French satellite technology experiment.
Prognoz	14 Apr 1972	A-2	Soviet solar studies.
Proton	16 Jul 1965	D	Soviet high-energy particle experiment.
Radarsat	1991	Shuttle	Canadian SAR remote sensor.
Radio	26 Oct 1978	F-2 (Piggyback)	Soviet ham radio satellite.
Raduga	22 Dec 1975	D-1	Soviet geosynchronous communications satellite.
Sakura/CS (Communications Satellite)	15 Dec 1977	Delta	Japanese experimental communications satellite.
San Marco	15 Dec 1967	Scout	Italian atmospheric research satellites.
Shinsei	28 Sep 1971	M-4S-3	Japanese cosmic ray measurement.
Signe	17 Jun 1977	C-1	French X-ray, gamma ray experiment.
Sirio	25 Aug 1977	Delta	Italian experimental communications satellite.
SPAS (Shuttle Pallet Satellite)	18 Jun 1983	Shuttle	W. German reusable experiment platform.
SPOT (Satellite Probatoire de l'Observation de la Terre)	22 Feb 1986	Ariane-1	European earth imaging system.

Spacecraft	1st Launch	Launch Vehicle	Mission
Sputnik	4 Oct 1957	Type A	1st artificial satellite (Soviet).
SRATS (Solar Radiation and Thermospheric Structure) satellite	24 Feb 1975	M-3C-2	Japanese solar radiation experiment.
Starlette	6 Feb 1975	Diamant B-P4	French geodetic satellite.
STW	8 Apr 1984	CZ-3	1st geostationary Chinese communications satellite.
Symphonie	19 Dec 1974	Thor Delta	French/W. German telecommunications satellite.
TDF/TV-SAT	1987	Ariane	French/W. German direct TV.
Telesat	4 Aug 1984	Ariane	French telecommunications.
Tele-X	1988	Ariane	Scandinavian direct TV/data.
Tenma/Astro	20 Feb 1983	M-3S-3	Japanese X-ray observations.
TOPEX (The Ocean Topography Experiment) satellite	1991	Ariane	U.S./French ocean altimetry.
Tournesol/Aura	15 Apr 1971	Diamant B	French solar radiation experiments.
UME/ISS (Ionospheric Sounding Satellite)	29 Feb 1976	N-1	Japanese ionospheric experiments.
Viking	22 Feb 1986	Ariane	Swedish space plasma experiments.
Yuri/BSE (Broadcasting Satellite Experiment)	8 Apr 1978	Delta	Japanese direct TV broadcast.

INDEX